Make: 3D Printing Projects

Brook Drumm, James Floyd Kelly,
Brian Roe, Steven Bolin, John Edgar Park,
John Baichtal, Rick Winscot, Nick Ernst,
and Caleb Cotter

MAKER MEDIA™
SAN FRANCISCO, CA

Make: 3D Printing Projects

by Brook Drumm, James Floyd Kelly, John Edgar Park, Brian Roe, John Baichtal, Rick Winscot, Steven Bolin, Nick Ernst, and Caleb Cotter

Printed in Canada.

Published by Maker Media, Inc., 1160 Battery Street East, Suite 125, San Francisco, CA 94111.

Maker Media books may be purchased for educational, business, or sales promotional use. Online editions are also available for most titles (*http://safaribooksonline.com*). For more information, contact our corporate/institutional sales department: 800-998-9938 or *corporate@oreilly.com*.

Editor: Anna Kaziunas France
Production Editor: Nicholas Adams
Copyeditor: Charles Roumeliotis
Proofreader: Kim Cofer
Indexer: Wendy Catalano

Interior Designer: David Futato and Anna Kaziunas France
Cover Designer: Anna Kaziunas France
Illustrator: Rebecca Demarest

October 2015:   First Edition

Revision History for the First Edition
2015-10-02:   First Release

See *http://oreilly.com/catalog/errata.csp?isbn=9781457187247* for release details.

978-1-4571-8724-7

[TI]

Contents

Preface

It's time to stop treating 3D printers like standalone machines—*they're just another tool in your toolbox.*

You're holding a book full of projects, ranging from simple and utilitarian to the utterly fantastical. While these projects require a 3D printer to produce custom parts, each utilizes a unique combination of electronics, hand assembly techniques, and additional tools, parts, and software. The printer is a custom fabrication enabler, but it's what you create with it that matters.

Because we've rejected the hype that puts the printer on a pedestal and embraced the "another tool" philosophy, we've found our printers to continue to prove themselves time-saving, multifaceted machines. By reducing the amount of time spent in front of traditional tools, they speed our making. Why drill a 1/4" hole in a small component when you can design the part with the hole already properly and accurately placed?

The additive process also saves us time, money, and pain. Hobbyists who use wood or aluminum for their mock-ups spend a good portion of their time measuring, cutting, and drilling. As mistakes are made or small modifications must be incorporated, they often have to start again from scratch. And again. And again.

This cycle is dramatically shortened when you design in software, test print, tweak, then reprint, as exemplified in Chapter 2 by the Raygun Pen project's pile of prototyped handles. A new piece can be printing while you work on another aspect of your project, or engage in something more mundane, like doing the dishes or mowing the lawn.

But a 3D printer can't do it all. If your project needs some LED lighting or motors added, those can't be printed up, wired, and soldered together to create a final product.

Perhaps you've had a 3D printer for some time now, or were looking for a reason to buy one, but the "Yoda head" figurines are no longer novel. Your machine may even be sitting in the corner, gathering dust. *We say it's time to put that printer back to work!*

Each design in this book was envisioned, lovingly prototyped, and assembled by a unique individual. Together, they exemplify the broad range of highly personalized, limit-pushing project possibilities enabled by the printing of parts, fusion of tools, and wide availability of affordable electronic components.

Even if you've never touched a printer, we hope these projects excite and embolden you to learn new skills, extend your current abilities, and awaken your creative impulses. If you're inspired to go make something after reading this book, then we've done our job.

—**Brook Drumm**, Founder and CEO of Printrbot

—**James Floyd Kelly**, Writer. Maker. Adventurer

WHAT YOU NEED TO KNOW

This book takes a step-by-step approach to projects, but it does assume that you understand how to solder and know a little about electronics. If these things are new to you, there are resources widely available on the Internet (we link to many of them in the projects) and Maker Media has published many books on these subjects.

PROJECT FILES AND CODE

Every project in this book was designed to fit on a Printrbot Simple Metal, with a build area size of 6"x6"x6".

PROJECT DOWNLOADS

The latest fabrication files, code, and resources for every project in this book can be found on the *Make: 3D Printing Projects* site (*https://github.com/ Make3DPrintingProjects*).

HOW THIS BOOK IS ORGANIZED

The projects in this book start off slowly and steadily become more complex.

Chapter 1: Lamp3D
Caleb Cotter kicks off the book with an awesome modular lamp project that's suitable for beginners and quickly gets you incorporating electronics into 3D-printed structures.

Chapter 2: 1950s Raygun Pen
Have you ever *really* liked something, but wanted it to be just a little different? James Floyd Kelly takes you step-by-step through a primer on how to augment an existing object through rapid prototyping.

Chapter 3: Two-Axis Camera Gimbal
You *could* pay through the nose for quality video stabilization, or you can 3D-print Nick Ernst's design to get your own camera gimbal for less than half of what it would cost to buy one.

Chapter 4: BubbleBot
John Baichtal's BubbleBot machine not only cranks out the bubbles, it has customizable speed settings for (at least) three different bubble styles.

Chapter 5: DDriver Rechargeable Screwdriver
Two short-lived battery-powered screwdrivers prompt Rick Winscot to begin a warranty-voiding teardown, resulting in a 3D-printed screwdriver redesign.

Chapter 6: Animatronic Eyes
Former special effects technician Brian Roe walks you through the process of creating a set of fully functional, extraordinarily expressive eyes that could easily have been built for a creature in a film.

Chapter 7: Inverted Trike RC
It's sleek, shiny, and fast, just like an expensive off-the-shelf RC car, but it's completely customizable. Steven Bolin's affordable design makes clever use of easily swappable 3D-printed replacement parts for those inevitable collisions.

Chapter 8: SkyCam
Brook Drumm's SkyCam is a little robot that travels on a rope or string, and can even go around corners! It's remote-controlled from your phone or web browser and it streams live video from its pan and tilt camera.

Chapter 9: Chauncey: The Wrylon Robotical Flower Care Robot
Meet Chauncey, a fully operational (and absolutely adorable) plant monitoring robot who charmingly waters your foliage with a tiny watering contraption. John Edgar Park takes you through every detail of the complex build.

ABOUT THE AUTHORS

Brook Drumm is the founder and CEO of Printrbot, Inc. Brook is an American maker who set out to start a side business in his garage. After a wildly successful Kickstarter in 2011, Brook was catapulted to the white-hot intersection of crowd funding, 3D printing, and the exploding maker culture. Printrbot is an example of what blood, sweat, and tears can produce if you set your mind and heart on what you are passionate about.

James Floyd Kelly is a writer who lives in Atlanta, GA with his wife and two young sons. He has degrees in Industrial Engineering and English and enjoys making things, writing about those things, and training young makers. He has written over 25 books on a variety of subjects from Lego robotics to open source software to building your own CNC machine and 3D printer.

John Edgar Park is a producer at Disney Research. He has worked in animation production at Disney for 10 years and has worked in computer graphics since 1994 at various companies, including IBM, Novalogic game studio, and Sony Pictures Imageworks. John was the host and co-writer of the Emmy Award-nominated series *Make: Television*. He regularly demos his projects at Maker Faires, and builds and writes about technology projects for *Make:* magazine and other places online and in print.

Brian Roe is a Tinkerer at heart and a Mechanical Designer by trade. His diverse career has led him down many interesting paths. He has worked as a creature creator for Hollywood films, and was part of a successful combat robot team for the show *Battlebots*. He also assisted with the design of the 3D cameras used for the film *Avatar*. Currently, he's an engineer at Printrbot and working on The 10,000 Year Clock project for The Long Now Foundation.

Steven Bolin is Printrbot's Production Manager and R&D department assistant. He has always enjoyed working with his hands, from construction to just projects around the house, and finds designing and building 3D printers an absolute joy. As a father of two kids and former Youth Pastor, he loves making and playing with toys and 3D printing naturally lends itself to this hobby. Steven hopes to continue developing new products that incorporate 3D printing into everyday life.

Rick Winscot has code in brain, soldering iron in hand, and Art Blakey blaring in the background. He transforms techno-babble into reality… and is strangely fond of the ellipsis.

John Baichtal writes books about toys, tools, robots, and hobby electronics. His first book was *The Cult of Lego*, an exploration of the culture and work of adult Lego builders. He's also the author of *Maker Pro*, *Robot Builder*, and *The Locksmith's Apprentice*, a fantasy novel. He lives in Minneapolis with his wife and three children.

Nick Ernst served in the Marine Corps for eight years, then decided to go back to school to pursue a career in Electronics Engineering. Before working on electronics at Printrbot Inc., he spent two years at Parallax Inc. as the lead developer of the Elev-8 multirotor platform. He's a maker at heart, and enjoys tinkering and hacking on all shapes and sizes of electronics. He believes that if it ain't broken, you haven't tinkered enough!

Caleb Cotter is an R&D Specialist at Printrbot. He's a maker with a passion for 3D printing and rapid prototyping, and is fascinated with the opportunities presented by desktop manufacturing. When he's not in his garage hacking—he's *still* in his garage wrenching on his car.

CONVENTIONS USED IN THIS BOOK

The following typographical conventions are used in this book:

Italic
> Indicates new terms, URLs, email addresses, filenames, and file extensions.

`Constant width`
> Used for program listings, as well as within paragraphs to refer to program elements such as variable or function names, databases, data types, environment variables, statements, and keywords.

`Constant width bold`
> Shows commands or other text that should be typed literally by the user.

 This element signifies a tip, suggestion, or general note.

 This element indicates a warning or caution.

USING CODE EXAMPLES

This book is here to help you get your job done. In general, you may use the code in this book in your programs and documentation. You do not need to contact us for permission unless you're reproducing a significant portion of the code. For example, writing a program that uses several chunks of code from this book does not require permission. Selling or distributing a CD-ROM of examples from Make: books does require permission. Answering a question by citing this book and quoting example code does not require permission. Incorporating a significant amount of example code from this book into your product's documentation does require permission.

We appreciate, but do not require, attribution. An attribution usually includes the title, author, publisher, and ISBN. For example: "*Make: 3D Printing Projects* by Brook Drumm and James Floyd Kelly (Make). Copyright 2016, 978-1-4571-8724-7."

If you feel your use of code examples falls outside fair use or the permission given here, feel free to contact us at *bookpermissions@makermedia.com*.

SAFARI® BOOKS ONLINE

 Safari Books Online is an on-demand digital library that delivers expert content in both book and video form from the world's leading authors in technology and business.

Technology professionals, software developers, web designers, and business and creative professionals use Safari Books Online as their primary resource for research, problem solving, learning, and certification training.

Safari Books Online offers a range of plans and pricing for enterprise, government, education, and individuals.

Members have access to thousands of books, training videos, and prepublication manuscripts in one fully searchable database from publishers like O'Reilly Media, Prentice Hall Professional, Addison-Wesley Professional, Microsoft Press, Sams, Que, Peachpit Press, Focal Press, Cisco Press, John Wiley & Sons, Syngress, Morgan Kaufmann, IBM Redbooks, Packt, Adobe Press, FT Press, Apress, Manning, New Riders, McGraw-Hill, Jones & Bartlett, Course Technology, and hundreds more. For more information about Safari Books Online, please visit us online.

HOW TO CONTACT US

Please address comments and questions concerning this book to the publisher:

Make:
1160 Battery Street East, Suite 125
San Francisco, CA 94111
877-306-6253 (in the United States or Canada)
707-639-1355 (international or local)

Make: unites, inspires, informs, and entertains a growing community of resourceful people who undertake amazing projects in their backyards, basements, and garages. Make: celebrates your right to tweak, hack, and bend any technology to your will. The Make: audience continues to be a growing culture and community that believes in bettering ourselves, our environment, our educational system—our entire world. This is much more than an audience, it's a worldwide movement that Make: is leading—we call it the Maker Movement.

For more information about Make:, visit us online:

Make: magazine: *http://makezine.com/magazine/*
Maker Faire: *http://makerfaire.com*
Makezine.com: *http://makezine.com*
Maker Shed: *http://makershed.com/*

We have a web page for this book, where we list errata, examples, and any additional information. You can access this page at *http://bit.ly/make3Dprintingprojects*.

To comment or ask technical questions about this book, send email to *bookquestions@oreilly.com*.

Lamp3D

by Caleb Cotter

Lamp3D is a small, repositionable, gooseneck-style LED lamp. It's a fun and easy introductory project for those just starting out with 3D printing, electronics, and soldering.

The minimalistic design was inspired by the Loc-Line modular hose/pipe system that's been used for anything from CNC machine lubrication to flexible shower heads.

All of the lamp's structural components are 3D-printed, with no additional hardware required. We can easily customize it by adding more linkages (or "vertebrae") or by modifying the lamp shade, and we can fabricate it on a very small print bed.

COST	+ $35
PRINT TIME	+ 6 hours
BED SIZE	+ 6"x6"x6"
ASSEMBLY	+ 30 minutes

PARTS, TOOLS, AND FILES

FILES TO DOWNLOAD

To complete the *Lamp3D* project, you'll need to download the fabrication files (*http://bit.ly/1RerzPq*) from the *Make: 3D Printing Projects* site (*https://github.com/Make3DPrintingProjects*).

PARTS

Description	Part Number
Wall adapter power supply - 9VDC 650mA	SparkFun TOL-00298
9V battery plug (optional)	SparkFun PRT-09518
Female DC barrel jack	SparkFun PRT-13126
FemptoBuck LED driver	SparkFun COM-12937
3-watt LED	SparkFun COM-13105
Red hookup wire	SparkFun PRT-08023
Black hookup wire	SparkFun PRT-08022

TOOLS & MATERIALS

3D printer	ABS filament
Soldering iron and solder	Heat gun or lighter
Wire strippers	Needle-nose pliers
Superglue	5/8" heat shrink tubing
Zip ties	1/4" heat shrink tubing

3D PRINTABLE FILES		
Quantity	**Description**	**Filename**
1	Lamp Base (ABS or PLA)	*Base.stl*
1	Lamp shade (ABS)	*LampShade.stl*
9+	Vertebrae (ABS)	*Vertebrae.stl*

GET YOUR PARTS PRINTING!

Print out all the files, one base, nine or more vertebrae, and one lamp shade.

I recommend printing these parts in ABS at 100% infill. PLA is less flexible and could snap when it comes time to put the pieces together.

While your printer is cranking out the plastic bits of goodness, it's time to get started on the lamp's electronics!

WIRING THE BARREL JACK

Cut two 2″ sections from each color of our hookup wire (one black and one red) (Figure 1-1).

Strip about 1/4″ of insulation from each end.

Unscrew the black housing on the barrel jack and solder the red wire to the center and the black to the outer leg. Then slide the 1/4″ of heat shrink tubing around the red wire connection to insulate it (Figure 1-2).

Figure 1-1 *Unstripped wires and barrel jack*

Use a pair of needle-nose pliers to close the collar around the wires, then screw the black housing back on as seen in Figure 1-3.

Figure 1-2 *Wires soldered to barrel jack*

Figure 1-3 *Barrel jack housing replaced*

ADD THE LED DRIVER TO THE BARREL JACK

Now it's time to add the FemptoBuck LED driver to the wired barrel jack you've just created.

Solder the black wire to PGND (marked on the board) and the red wire to VIN (Figure 1-4).

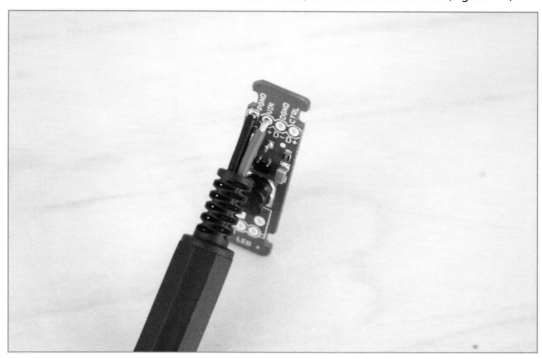

Figure 1-4 *Solder wires to driver board*

Take a zip tie, wrap it around the wires, and seat it in the notches on the sides of the Femp-toBuck (Figure 1-5). This will provide strain relief for our solder joints.

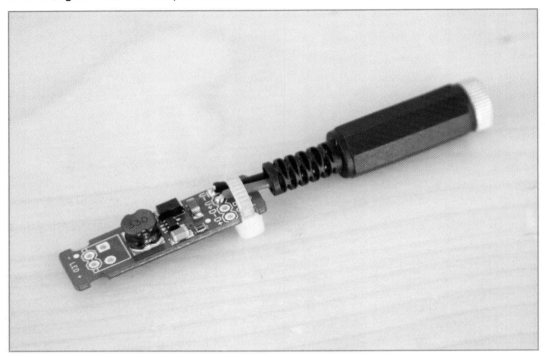

Figure 1-5 *Driver zip-tied to jack*

Now set this assembly aside as we add some long wires to our LED.

SOLDER WIRES TO THE LED

Cut off 12" of the red and black hookup wires and strip back about a 1/4" of insulation off of each end (Figure 1-6).

Note the positive (**+**) and negative (**-**) markings on the LED board.

The wires soldered to the LED need to reach all the way through the neck of your lamp. It's safest to make them longer than you actually need, especially if you're printing your parts as you're assembling the electronics.

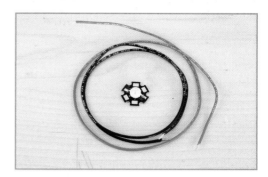
Figure 1-6 *12" of wire and LED*

Solder the red wire to the pad next to the positive indicator and the black wire next to the negative one as shown in Figure 1-7.

Now go make a sandwich (or something else from this book) while you wait for the prints to finish.

Figure 1-7 *Soldered LED connections*

ASSEMBLE THE ARM

I hope your sandwich was tasty.

Now that your prints are finished it's time to put this thing together!

Start by grabbing the LED assembly (Figure 1-8) and feeding the wires through the lamp shade. Glue the LED down in the center.

Take the vertebrae and firmly snap them into one another. Then slide the assembled vertebrae over the wires and snap in the lamp shade as shown in Figure 1-9.

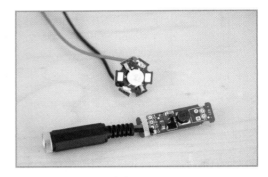

Figure 1-8 *Soldered electronics*

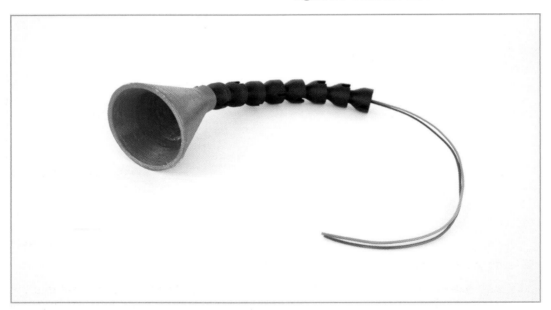

Figure 1-9 *Assembled lamp arm*

CONNECT THE BASE

To mate up the base with the arm, feed the wires from the arm down through the base as shown in Figure 1-10 and snap the arm into the base.

Believe it or not we are almost done!

Figure 1-10 *Arm connected to base*

FINAL ELECTRONICS ASSEMBLY

To finish up, grab the LED driver assembly we set aside earlier.

Take the LED wires protruding from the base, trim them, and solder the red wire into the positive (**+**) LED hole on the FemptoBuck; solder the black wire into the negative LED (**-**) hole (Figure 1-11).

Figure 1-11 *Driver soldered to LED wires*

Zip-tie the wires, seating the zip tie into the notches for strain relief, just like you did for the barrel jack assembly.

Before you seal up the electronics, test the lamp by plugging in the 9V wall wart (Figure 1-12). Wait for the moment of truth. If it works—congratulations!

If the lamp doesn't light up, go through the "Troubleshooting" box before moving on.

Figure 1-12 *The moment of truth!*

TROUBLESHOOTING

If your light isn't turning on, here are a few things to try:

1. Go back and check all of your instructions and make sure your connections are all in the right places.

2. Take a close look at all of the solder joints and make sure there are no cold solder joints. Cold solder joints are dull and can look like they are resting on the pad instead of adhered to it. A good solder joint will be shiny and seem to flow over the connection.

FINISHING UP

To complete the project, slide the 5/8" heat shrink tubing over the plug (see Figure 1-13) and LED driver assembly and use a heat gun or lighter to shrink it down.

Figure 1-13 *Heat-shrink covered connections*

Congratulations! You have successfully built your very own Lamp3D (Figure 1-14)!

Use it as a work light to build more electronics or get creative with some of the upgrades and improvements listed in the next section.

POSSIBLE UPGRADES AND IMPROVEMENTS

Here are some possible Lamp3D upgrades you could try, but feel free to get creative and create your own unique modification.

On the aesthetic side, you could modify the lampshade to add lenses and filters, or shape it differently to create some cool effect. You could also modify the lamp base so that the electronics are tucked inside the enclosure.

If you liked assembling the electronics part of this project, why not take it further by adding an Arduino to the DGND and CTRL pins and using PWM to adjust the brightness?

Plus, you could always add lasers—because who doesn't love lasers?

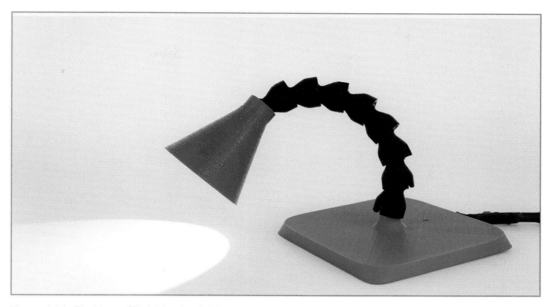

Figure 1-14 *Final Lamp3D shining brightly!*

1950s Raygun Pen

by James Floyd Kelly

A few years back my wife gave me a nice pen that rested on a grip-shaped base as a fun little birthday gift for my home office.

Recently I began to wonder if it might be possible to add a special effect or two to the pen and base. After all, the pen looked just like a laser pistol right out of a science fiction movie! Pew Pew!

After examining the pen and base, I determined that the base wasn't hollow and was too small to really allow me to add any electronics inside. So, I decided to start from scratch.

In this project I'll show you how I used my 3D printer to create custom parts to retrofit an existing object and I'll pass along a few prototyping tricks that you might find useful in your own projects.

By the end of the chapter, you'll be able to duplicate the work I've done to create your own 1950s Style Raygun Pen—or have found a way to improve it!

COST
+ $50

PRINT TIME
+ 3 hours

BED SIZE
+ 4"x4"x4"

ASSEMBLY
+ 2 hours

PARTS, TOOLS, AND FILES

PARTS

Description	Part number
Trinket, mini-microcontroller 5V logic	Adafruit 1501
NeoPixel Ring 16xWS2812 5050 RGB LED with integrated drivers	Adafruit 1463
3xAAA battery holder with on/off Switch and 2-pin JST	Adafruit 727
Blank mint tin	Adafruit 97
5" female jumpers (10 pack) with 40 headers	Schmartboard (*http://www.schmartboard.com*) 920-0006-01
7" blue female jumpers (10 pack) with 40 headers	Schmartboard (*http://www.schmartboard.com*) 920-0007-01

TOOLS & MATERIALS

3D printer	Silver PLA filament	Red PLA filament
Soldering iron and solder	Breadboard	Hot glue gun and glue
Calipers	Drill press or Dremel rotary tool	Wire snips
Pliers	Electrical tape	Spray paint

3D PRINTABLE FILES		
Quantity	**Description**	**Filename**
1	Attaches the NeoPixel Ring to the black mint tin	*led_housing.stl*
1	Lefthand side of the Raygun grip	*left_grip_final.stl*
1	Righthand side of the Raygun grip	*right_grip_final.stl*

BRAINSTORMING THE RAYGUN PEN

Every good retrofitting project begins with a brainstorming session.

I already knew that I wanted my Raygun Pen to "do something" other than just allow me to write or draw. Lots of ideas popped into my head.

The Raygun Pen could:

- Make a "pew pew" sound when a button is pressed
- Put some LEDs in the grip that light up when the pen is removed
- Play a theremin-like space sound when the pen is placed on the base
- Spin 360 degrees while the pen is resting in the base

After looking at various electronic components, I ultimately settled on using a tiny micro-controller board called a *Trinket* to control a *NeoPixel Ring*. The project is powered by a tiny battery box that holds three AAA batteries and provides the 5V needed by the electronics.

In the end, I settled on a simple short-term goal for the Raygun Pen. When the user places the pen on the base, which is shaped like the grip of a raygun, it closes a circuit. This causes the LEDs in the base to light up and cycle through a fun light pattern. When the pen is lifted from the base, the circuit is broken and the lights turn off.

SELECTING A PEN

Before you begin constructing your Raygun Pen, you're going to need to purchase the actual pen that will sit on the base.

There are easily hundreds of suitable pens out there. When I began hunting for a pen to use, I wanted something that had a distinctive look to it—something reminiscent of early science fiction movies and those crazy-looking rayguns that the spacemen would fire. What I chose was the Cross Edge Nitro Blue (Figure 2-1)—normally I would never have selected a $40 pen but I got it on sale for $15 and it was perfect.

Figure 2-1 *Cross Edge Nitro Blue pen*

Ultimately, you're going to want to pick one that suits your tastes and the type of device you wish to build. But before you make a purchase, find a pen that has an outer metal shell (not plastic) or at least a metal clip.

Why a metal body or metal clip? The metal will be used to help close the circuit that will be built inside the base and grip of the Raygun Pen.

 You don't have to have a metal pen or metal clip to close the circuit. You could easily incorporate a simple on/off switch on the base that turns the NeoPixel Ring animation on and off.

PROTOTYPING THE GRIP

After selecting the pen, the next step is to create the shape of the grip that will hold the pen.

You could use a ruler or calipers to take measurements of your pen and then create a grip using a CAD application—at first that's exactly what I intended to do. My original grip, however, was too angular—lots of sharp edges that made it look funny.

I decided to try and create some sketches of different grips. I found the easiest way to find something I liked was to make a few photocopies of my pen and then draw the grip on the photocopy, as seen in Figure 2-2.

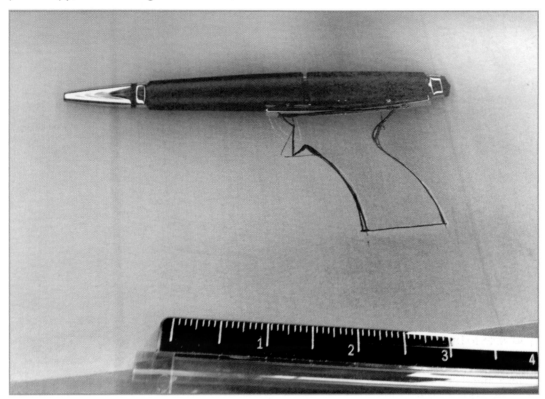

Figure 2-2 *Grip sketch on pen photocopy*

PAPER SKETCH TO SVG CONVERSION

After I dialed in the grip's look, I designed it in Tinkercad. Tinkercad allows you to use primitive objects (such as cubes, spheres, and pyramids) to create more advanced shapes, but it can be a very time consuming to create something like the shape shown in Figure 2-2.

Luckily, there's actually a much faster way to take a unique shape or sketch you've created and bring it into a CAD application.

Tinkercad can only import STL and SVG files, so for me to use my custom shape I'll need to convert it to the SVG file format. Fortunately, there's a free and easy online tool that can do this for you.

First, create a clean trace of the shape on a white piece of paper and then fill it in with a black marker. You'll end up with something that looks like Figure 2-3.

Take a photo or scan the image and save the JPEG image to your computer and then point your web browser to online-convert.com. Click the "Go" button in the Image Converter box and select the "Convert to SVG" option as shown in Figure 2-4.

A new screen will appear like the one in Figure 2-5. Click the "Choose Files" button and browse to the location of the JPEG image of your shaded shape. Also click the Monochrome option in the Color section. Click the "Convert file" button and your SVG file will be created.

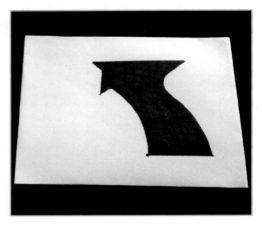

Figure 2-3 *Shaded grip sketch*

Figure 2-4 *Online-Convert.com*

IMPORTING INTO TINKERCAD

Next, this SVG file must be imported into Tinkercad. Log in to tinkercad.com and click the "Create new design" button shown in Figure 2-6.

Figure 2-5 *Convert image to SVG*

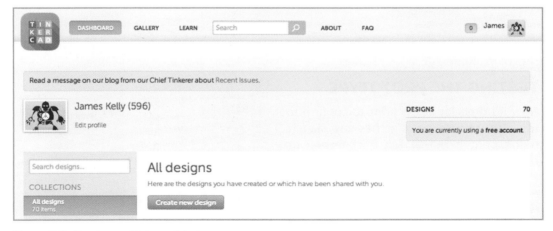

Figure 2-6 *Create new Tinkercad design*

On the right side of the screen, open the Import section. Click the "Choose File" button, browse to your new SVG file, and then click the "Import" button.

As you can see in Figure 2-7 the grip has been imported, but it is a bit large compared to my sketch measurements.

I knew what the length and width of the grip should be because I had measured the paper sketch shown in Figure 2-2.

RESCALING AND ADJUSTING THE IMPORTED SVG

I resized the imported SVG in Tinkercad by holding down the Shift key and then dragging one of the corner controls to lock in the length and width ratio.

Then I needed to adjust the thickness. My intent was to split the grip into two pieces that will be hollowed out, so I could route the wires used in the circuit.

Since Tinkercad allows me to copy a 3D model and flip or mirror the model, I changed the grip's thickness to half the final thickness.

Figure 2-7 *File imported too large*

 Tinkercad has plenty of tutorials that can help you understand the controls and features better.

After resizing the grip to the proper measurements, I ended up with the solid object shown in Figure 2-8.

TESTING THE PROTOTYPE

Now it's time for a test print to confirm that the size and shape matches my sketch. I didn't print it out at full thickness—I only need a few layers to confirm that the shape is a match. Figure 2-9 shows my test grip printed out in white filament. I was quite happy with it.

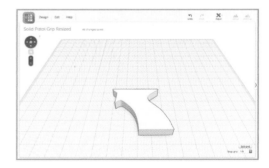

Figure 2-8 *Solid half-thickness Raygun grip*

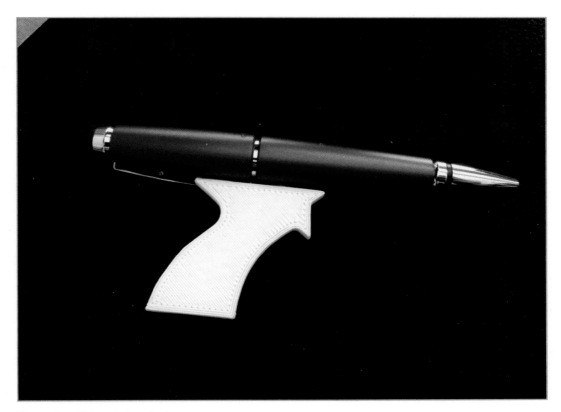

Figure 2-9 *Grip prototype test*

Next, back in Tinkercad, I used a mix of rectangles and triangles converted into "hole objects" (with the Hole button) to make room for the necessary wiring, as shown in Figure 2-10.

After merging the solid and hole objects, I ended up with a grip piece that was hollowed out. Figure 2-11 shows one-half of the grip.

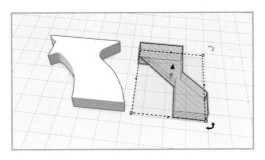

Figure 2-10 *Converted to a hole object*

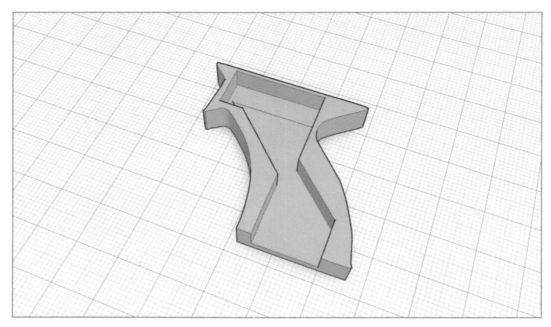

Figure 2-11 *Hollow grip*

I could have easily printed the entire grip out, but to save time, I only printed the parts that I needed to test for accuracy. I needed to insert the pen into the grip to ensure that the pen sat properly and that the metal clip could contact with the circuit-closing wires.

With Tinkercad, I can make copies of existing models and then delete away sections I'm not interested in printing. To test the grip, I deleted some sections, leaving only the top third for printing.

Over a series of test prints, some of which are shown in Figure 2-12, I was able to add in a hollow space at the top of each grip half that both allowed the pen to rest properly *and* enabled the pen clip to go far enough down into the grip to touch two wires.

Once I was happy with the grip, I printed out the two halves, then put them together underneath the pen to see what it looked like. Figure 2-13 shows the two gray halves unassembled and Figure 2-14 shows the assembled grip.

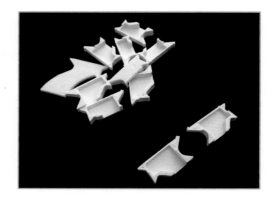

Figure 2-12 *A plethora of prototypes*

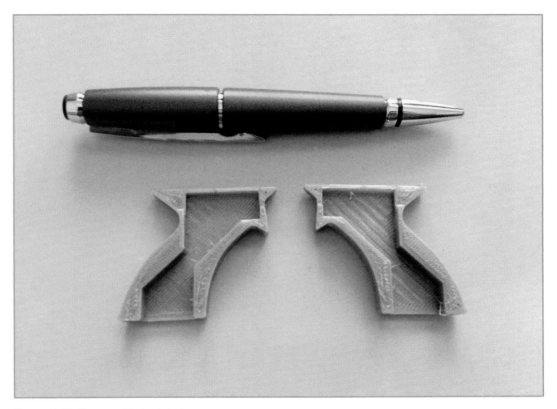

Figure 2-13 *Unassembled grip halves*

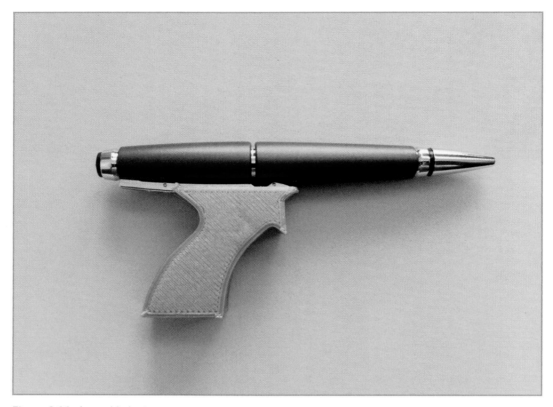

Figure 2-14 *Assembled grip*

DEVELOPING THE NEOPIXEL SHELL

The pen and grip will sit on top of a base that should be large enough to hold the battery box and the Trinket plus some wires. I went with one of the most popular and easily recognizable enclosures available to makers around the world—the mint tin. You can buy these from many sources or just find one in the grocery store.

 Mint tins make great project enclosures; they're cheap, strong, and open and close easily—and once closed the lid tends to stay closed. Plus, if you mess one up, it's easily replaceable.

I wanted the NeoPixel Ring to sit inside a small partially curved surface. The best way to model this in Tinkercad is to create a sphere and cut off and keep the very top portion while discarding the bottom. The piece that remains behind will be flat on the bottom and curved on top.

Since I wanted to convert my cut sphere into a shell for the NeoPixel Ring, I took the measurments of the ring's inner and outer diameters. Back in Tinkercad, I made a washer-like shape and converted it into a hole object that would be merged with the sphere. You can see the sphere fragment and the hole object in Figure 2-15.

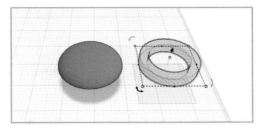

Figure 2-15 *Sphere fragment, hole object*

Next, I merged the ring hole object (Figure 2-16) and the sphere fragment and ended up with the NeoPixel Shell shown in Figure 2-17.

Due to the too-tight fitting mishaps that I relate in "Prototyping Pitfalls", I added a number of hole objects to the NeoPixel Shell so I could not only pop the ring out, but also so I would be able to route wires from the Trinket to the ring to make it function.

The final NeoPixel Shell model is shown in Figure 2-18 and Figure 2-19. I printed it in Rocket Red!

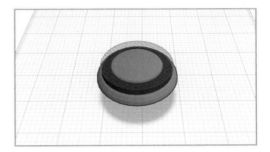

Figure 2-16 *Hole object and sphere overlay*

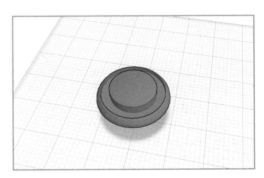

Figure 2-17 *Final merged NeoPixel Shell*

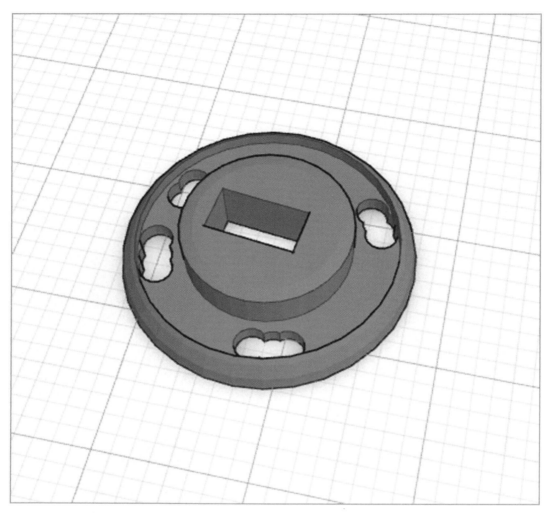

Figure 2-18 *Final Tinkercad model*

PROTOTYPING PITFALLS

Funny story—it took three test prints to fine-tune the outer and inner dimensions of the hole ring before the NeoPixel Ring snapped down into the shell. And that's when I figured out my mistake. The NeoPixel Ring was such a snug fit I couldn't get the ring out! I couldn't get a good grip on the ring with tweezers and I didn't want to risk damaging the NeoPixel Ring, so after carefully drilling a few small holes on the back of the shell I was able to push it out.

Figure 2-19 *Final Rocket Red printed base*

That's it for the 3D printer! But the Raygun Pen isn't quite done yet…

PROTOTYPING THE CIRCUIT

Ask anyone who has done any type of electronics prototyping for any length of time, and they'll tell you that it's always a smart idea to test your circuits first before transferring them into an enclosure or soldering more permanent connections.

I absolutely love to use the female jumpers from Schmartboard (also available at select RadioShack stores)—these wires are very flexible and they make it super easy to plug and unplug connections quickly during prototyping.

SOLDER THE PCB HEADERS

The Trinket and NeoPixel Ring connections are at fixed locations on the printed circuit boards (PCBs), so I recommend that you solder headers to the Trinket and the NeoPixel ring (as shown in Figure 2-20) and use female jumpers to connect the components.

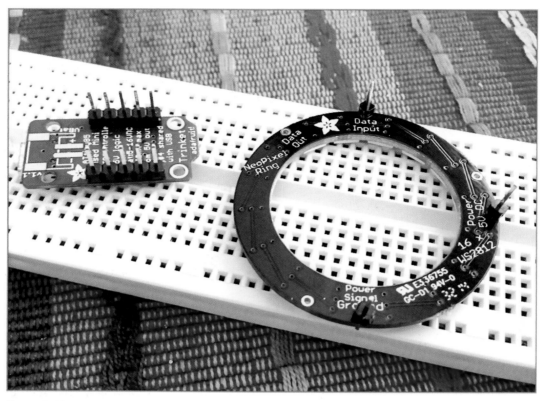

Figure 2-20 *Trinket and NeoPixel with soldered headers*

Trinket headers

    The Trinket arrives from Adafruit with precut headers in the package; solder them to the PCB. Later, you'll be attaching jumpers to the connections labeled #0, BAT, and GND.

NeoPixel Ring headers

    Use the headers included with the jumper pack. Snap off three single headers with pliers and solder one to each of the Data Input, Power 5V, and Power Signal Ground holes.

 Optionally, you could forgo the headers and solder wires directly to PCBs.

TRINKET SETUP TRICKS

In order to use the Trinket, you'll need a to install a modified version of the Arduino IDE. There are a few tricky bits, so read through the Adafruit Introducing Trinket Guide (*http://bit.ly/1P5wJhv*) to set up and learn to use your Trinket.

After you've gone through the Trinket Guide, connect your Trinket to your computer via USB cable and run the Blink sketch. You can locate this example program from File menu→Examples→01.Basics→Blink. When the tiny LED on the board blinks, you'll know that the hardware is functioning properly.

NEOPIXEL ANIMATION

Once you're sure you have a working Trinket, it's time to test the NeoPixel ring and tweak the animation code.

DOWNLOAD THE CODE

Head to the Adafruit website to download the NeoPixel library (*http://bit.ly/1KLaASC*) and while you're there, take a look at the original version (*http://bit.ly/1O7VM4m*) of the NeoPixel Ring program I've modified for this project.

If you haven't downloaded my code (*raygun_blue_spin_final.ino* (*http://bit.ly/1L43j1D*)) for this project yet, grab that as well.

INSTALL THE NEOPIXEL LIBRARY

Next, you'll need to install the NeoPixel library in order to run the Arduino code. You can do this easily with the built-in library manager (Arduino Sketch menu→Include Library→Manage Libraries); the NeoPixel library is already present.

UNFAMILIAR WITH ARDUINO LIBRARIES?

Installing Arduino libraries (like the Adafruit NeoPixel library you'll need for this project) can be confusing for new users. Since several projects in this book use Arduino libraries, we've added step-by-step instructions in Appendix A.

The original NeoPixel Ring program does a number of things. It cycles through multiple animation pattern demos that consist of three colors—red, blue, and green.

I decided I liked the pattern where a single LED lights up at a time, racing around the perimeter. I slightly modified the original Arduino program by removing the bits I didn't need and changing the LED color to blue in the *raygun_blue_spin_final.ino* file, also shown in the following code example.

CHANGING THE LED COLOR

The following code example uses hexidecimal notation to set the colors: `0xff0000` for red, `0x0000ff` for blue, and `0x00ff00` for green. To change colors, swap out the line **`uint32_t color = 0x0000FF;`** for another hex color.

NEOPIXEL RING CODE

```
// Raygun Pen program
// Modified the original version from Adafruit Goggles project

#include <Adafruit_NeoPixel.h>
#ifdef __AVR_ATtiny85__ // Trinket, Gemma, etc.
#include <avr/power.h>
#endif

#define PIN 0

Adafruit_NeoPixel pixels = Adafruit_NeoPixel(32, PIN);

uint8_t  mode   = 0, // Current animation effect
         offset = 0; // Position of spinning LED
uint32_t color  = 0x0000FF; // Start blue
uint32_t prevTime;

void setup() {
#ifdef __AVR_ATtiny85__ // Trinket
  if (F_CPU == 16000000) clock_prescale_set(clock_div_1);
#endif
  pixels.begin();
  pixels.setBrightness(85); // 1/3 brightness
  prevTime = millis();
}

void loop() {
  uint8_t  i;
  uint32_t t;
  {
    for (i = 0; i < 16; i++) {
      uint32_t c = 0;
      if (((offset + i) & 7) < 2) c = color; // 4 pixels on...
      pixels.setPixelColor(  i, c); // NeoPixel in Shell
    }
    pixels.show();
    offset++;
```

```
        delay(50);
    }
}
```

CONNECT THE NEOPIXEL RING

Now that you've configured the animation code, you'll need to test it out on the NeoPixel Ring.

Connect the Trinket to your computer with the USB cable. Then use jumpers to connect the NeoPixel Ring connections to the Trinket as shown in Table 2-1.

Table 2-1 *Trinket connections*

TRINKET CONNECTIONS		
Trinket pin #0	→	NeoPixel `Data Input`
Trinket pin BAT	→	NeoPixel `Power 5V DC`
Trinket pin GND	→	NeoPixel `Power Signal Ground`

Load the code on the Trinket by clicking the upload button (right-pointing arrow) in the Arduino IDE.

Although the IDE shows the code has uploaded, it will appear that nothing has happened, but that's not the case. You can't power your NeoPixel Ring from your computer; it needs to be hooked up to a separate battery before it will illuminate.

Next you'll breadboard the circuit, connect a battery pack, and set the NeoPixels aglow!

BREADBOARD THE CIRCUIT

ATTACH JUMPERS TO THE BATTERY BOX

The battery box provides both V + and GND connections to the LED ring and the microcontroller board.

To make connecting and disconnecting the battery box much easier during testing, cut off one end of two female/female jumpers. Then clip the red and black wire exiting the battery box and solder on two jumper wires, as shown in Figure 2-21.

POWER, GROUND AND PCB CONNECTIONS

In this next step, you'll connect the Trinket, NeoPixel Ring, and battery box using jumpers and a breadboard.

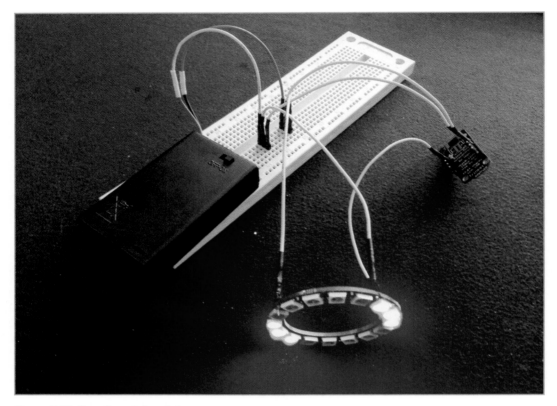

Figure 2-21 *Breadboarded prototype*

Because this project uses female/female jumpers, cut two rows of 3-pin headers and insert them into the breadboard, then connect the female jumpers to the headers.

 The breadboard and PCB connections are described in detail in Table 2-2.

As the breadboard setup shows in Figure 2-21, now that it's finally connected to the battery pack, the NeoPixel Ring is displaying the blue "racing" pattern.

Table 2-2 *Breadboarded connections, Figure 2-21*

BREADBOARD CONNECTIONS		
Voltage header row		
1	→	RED battery wire to yellow jumper
2	→	NeoPixel Power 5V DC
3	→	Trinket BAT
Ground header row		
1	→	BLACK battery wire to blue jumper
2	→	Trinket GND
3	→	NeoPixel Power Signal Ground
Trinket/NeoPixel connection		
Trinket pin #0	→	NeoPixel Data Input

 Breadboard connections are numbered left to right for each header row, as shown in Figure 2-21.

PEN TEST

Next, test out using the pen to close the circuit and activate the NeoPixel Ring. For the circuit to be closed by pen contact, it needs to be open by default.

First, insert a 2-pin header into the breadboard. Then disconnect the black/blue battery ground wire from the 3-pin header and attach it to the 2-pin header, as shown in Figure 2-22.

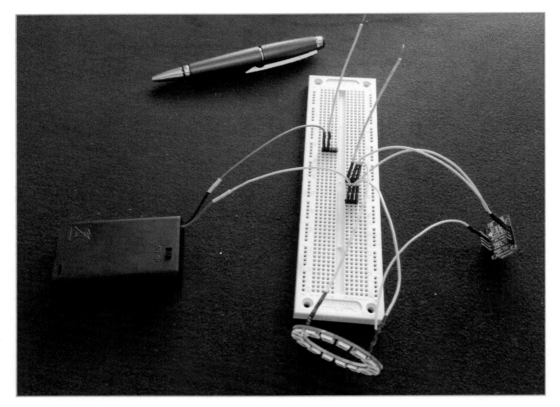

Figure 2-22 *Pen contact wires*

Cut and strip the ends of two "pen jumpers," providing contacts for the pen's metal clip to close the circuit. Attach one stripped jumper to the 2-pin header and the other to the former ground header row. Figure 2-22 shows the pen wires added to the existing circuit, sticking straight up, breaking the ground connection.

Now the moment of truth! Holding the nonmetalic parts of the pen, use the metal clip to bridge the two wires that are standing straight up.

If the NeoPixel Ring lights up, you've done it! If your prototype isn't successful, check your connections until you find the problem. Jumpers are very handy for troubleshooting connection problems; the connections are easily rearranged—no desoldering required.

TRANSFERRING THE CIRCUIT

Now that you've got a working circuit, it's time to transfer it from the breadboard to our final project enclosure.

HARDWIRE THE HEADERS

On the breadboarded circuit, the headers shared a row and the breadboard's internal connections created a common contact point. Now that you've removed the circuit from the breadboard, you'll need to re-create these shared connections.

Since you're replicating the final circuit shown in Figure 2-22, you'll need two 3-pin headers and one 2-pin header (Figure 2-23).

The easiest way I've found to create a single connection between attached headers is to strip the end from a piece of wire, then lay it across the headers. Add a blob of solder to combine the two as shown in Figure 2-24. Then clip off the excess wire protruding from the side.

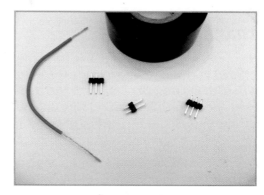

 You can also use braided wire twisted around the headers as a base for the solder to adhere to.

Figure 2-23 *Headers and stripped wire*

Wrap the soldered ends of the headers with some electrical tape as shown in Figure 2-25 to prevent any shorts from occurring; remember, you'll be stuffing these wired components into a small tin case, so use electrical tape to insulate any exposed leads whenever possible.

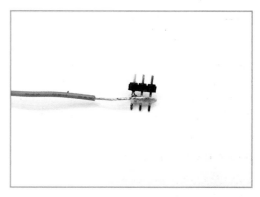

Figure 2-24 *Solder a wire across all three headers*

SPRAY-PAINT THE MINT TIN

Painting the tin will be easier if you remove the lid. Use a pair of pliers to gently pull apart the metal tabs attaching the lid until you have enough space to separate the two pieces. Then it's easy to paint the outer surface any color you like; I chose black as you can see in Figure 2-26.

TRACE THE BASE

Next, on a blank piece of paper, trace the 3D-printed base. Be sure to include the holes for where the wires will connect to the NeoPixel Ring and a hole for inserting wire up into the hollow grip. Figure 2-27 shows my tracing.

Tape or glue down the tracing to the *underside* of the tin lid. Orient the tracing so the Raygun Pen will point over the length of the base (otherwise it will tip over). The back of the base should be closest to the tin edge (see Figure 2-28 for general orientation), although that's jumping ahead a step.

CUT THE WIRING HOLES

Next, using your pattern as a guide, cut holes in the tin lid with a drill press or rotary tool. The circuitry inside the tin needs to pass through the 3D-printed base to reach the NeoPixel Ring and pen connection points.

Before connecting any wires, test-fit the Neo-Pixel Ring into the red base, and the base into the tin, to ensure that they fit properly, as shown in Figure 2-28.

Figure 2-25 *Taped, soldered headers*

Figure 2-26 *Spray-painted tin*

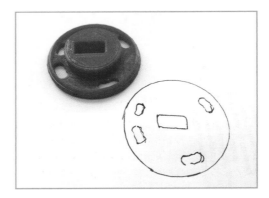

Figure 2-27 *Traced base*

Figure 2-28 *LED ring in 3D-printed shell on base*

While test fitting, double-check that the `Data Input`, `Power 5VDC`, and `Power Signal Ground` headers on the NeoPixel Ring are visible through the underside of the tin's lid.

Once you're satisfied with the holes, don't reattach the tin lid. It's easier to assemble the electronics with the two pieces still disconnected.

BEGIN INSERTING COMPONENTS INTO THE TIN

It's time to start fitting things into the tin. Your battery box may be slightly larger or smaller than mine, but you want to make certain it's placed where it won't block the Trinket, or the NeoPixel Ring headers that protrude through the lid and down into the tin.

Figure 2-29 shows the initial test fit that will allow you to close everything up and tuck all the wires in. Use blue painter's tape to hold things in place during test fitting. Once you're confident in the component placement, use some double-sided tape to secure the battery box and a small dab of hot glue to hold the Trinket in place.

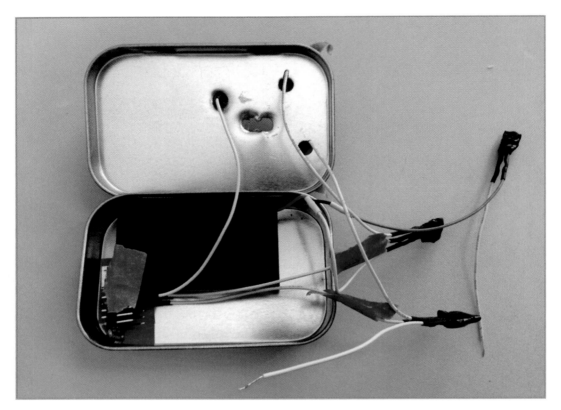

Figure 2-29 *Initial test enclosure fit*

FINAL ELECTRONICS ASSEMBLY

Table 2-3 *Final connections, as shown in Figure 2-29*

FINAL CONNECTIONS		
Voltage three-header row		
1	→	RED battery wire
2	→	NeoPixel Power 5V DC
3	→	Trinket BAT
Ground three-header row		
1	→	Pen Connection 1 (stripped end)
2	→	NeoPixel Power Signal Ground
3	→	Trinket GND
Pen two-header row		
1	→	BLACK battery wire
2	→	Pen Connection 2 (stripped end)
Trinket/NeoPixel connection		
Trinket pin #0	→	NeoPixel Data Input

Thread the *Pen Connection 1* and *Pen Connection 2* wires up through the largest hole in the tin lid and through the printed base as shown in Figure 2-30.

After the wires have been threaded through, reconnect the pieces of the painted mint tin and close the lid, securing the components inside.

 Turn on the battery box before closing the tin or the pen won't have a powered circuit to close.

Figure 2-30 *Protruding pen connections*

ADD THE GRIP

If you haven't already, glue the two halves of the gray grip together with hot glue.

Trim and strip the ends of the protruding jumpers and solder a single pin header to both; this makes them a bit easier to grab and thread through the hollow grip as you fit the grip into the red base, securing it with hot glue.

Next, clip the headers off of the "grip wires" and apply a bit of solder to the wire, *tinning* them to make them stiff. I folded them over the top of one side of the grip as shown in Figure 2-31.

As you prototyped previously, these "grip wires" form the open circuit that's closed when the pen clip touches both of the exposed ends.

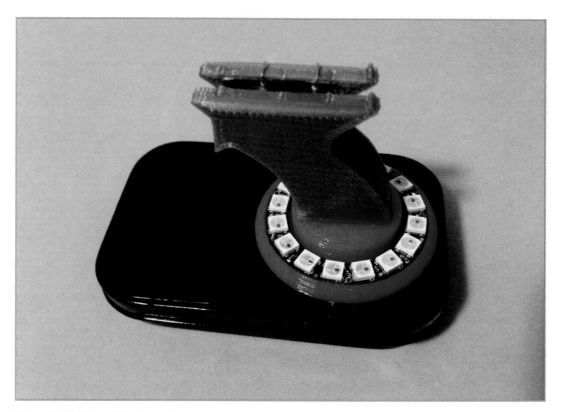

Figure 2-31 *Wires folded over grip*

LIGHT IT UP!

With the battery box turned on inside the enclosure, place the pen on the grip and hold your breath. A few seconds later, the LED ring should begin animating (see Figure 2-32).

Figure 2-32 *Finished Raygun Pen*

If removing the pen breaks the connection and the LEDs stop, it works!

Congratulations! You now have a working Raygun Pen, with the pen grip illuminated with blue LED animation.

UPGRADES AND IMPROVEMENTS

Although my Raygun Pen build worked just as I expected, I immediately began thinking of ways it could be improved:

Augment the grip files

> The exposed wires on the top of the grip work fine, but the design could be improved by creating two small holes in the grip's 3D model that would be used to hold the exposed wires. A dot of solder could be applied to each wire, keeping them from falling down inside the grip and ensuring proper contact with the pen's clip.

Skip the glue

> Instead of gluing the grip to the top of the shell, it would be preferable to create a clip or pin locking system to keep the NeoPixel Ring shell and grip halves securely in place.

Improve the grip's pen fit

> The grip has a flat surface and ideally it would be more rounded to "hug" the contour of the pen.

Add sound

> It would be super fun to have the LED ring light up and hear a PEW! PEW! sound go off when the pen is placed on the grip.

A custom case is probably in order, but check around to see what kinds of noisemaking options are available and whether they could be squeezed into the tin.

If you're new to 3D printing (and maybe even electronics), I hope my project has inspired you to engage in the world of possibilities that exist for creating your own custom projects. For me, a project like this one is never finished, so it's back to Tinkercad to begin as I develop additional improvements for Raygun 2.0!

Two-Axis Camera Gimbal

by Nick Ernst

Whether you're an avid quadcopter pilot, or you prefer to have your hobbies grounded, we all appreciate a steady picture from our camera.

With how much brushless motors have improved, it's no wonder that they've been incorporated into video stabilization. However, although these motors are widely available, a good camera gimbal is still a pricey purchase.

This project will allow you to 3D-print a quality camera gimbal for less than half of what it would cost to buy one. With minimal hardware and only the necessary electronics, this project is perfect to get you started with camera stabilization.

COST
+ $130

PRINT TIME
+ 12 hours

BED SIZE
+ 6"x6"x6"

ASSEMBLY
+ 2 hours

PARTS, TOOLS, AND FILES

FILES TO DOWNLOAD

To complete the *Two-Axis Camera Gimbal* project, you'll need to download the fabrication files and code (*http://bit.ly/1M05W0G*) from the *Make: 3D Printing Projects* site (*https://github.com/Make3DPrintingProjects*).

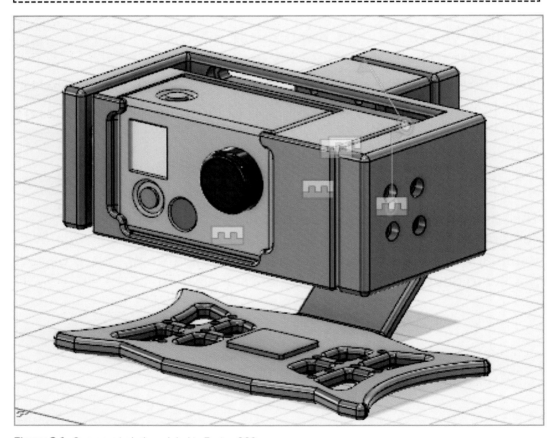

Figure 3-1 *Camera gimbal modeled in Fusion 360*

PARTS

Quantity	Part description	Part number
1	Quanum Micro AlexMos Brushless Gimbal Controller 2-Axis Kit Basecam (SimpleBGC)	HobbyKing PN: 9154000013-0
2	Turnigy HD 2212 brushless gimbal motors	HobbyKing PN: 9244000016-0
1	Turnigy 3-cell LiPo battery	HobbyKing PN: T1450-TX-3
1	624zz bearing	McMaster-Carr PN: 7804K103
8	M2 x 6mm flat head socket cap screws	McMaster-Carr PN: 91294A004
8	M3 x 6mm socket head cap screws	McMaster-Carr PN: 95263A110
4	M3 x 16mm socket head cap screws	McMaster-Carr PN: 95263A158
1	M4 x 12mm socket head cap screw	McMaster-Carr PN: 90128A214

 The Quanum Micro AlexMos Brushless Gimbal Controller 2-Axis Kit is also known as the SimpleBGC 8-bit. BaseCam Electronics renamed it when it introduced a 32-bit version. BaseCam maintains partner and store listings (https://www.basecamelectronics.com/wheretobuy).

TOOLS & MATERIALS

3D printer, printer filament	1.3mm allen wrench
2.5mm allen wrench	3.5mm allen wrench
Soldering iron and solder	Safety glasses
Double-sided tape	Micro USB cable

3D PRINTABLE FILES

Quantity	Description	Filename
1	Tilt Body	*Tilt-Body.stl*
1	Roll Body	*Roll-Body.stl*

3D PRINTABLE FILES		
1	Roll Motor Mount	*Roll-Motor-Mount.stl*
1	Base	*Base.stl*

TILT BODY ASSEMBLY

The Tilt Body holds your camera and allows the unit to compensate for movement on the x-axis, the "up and down" movement from the camera's perspective.

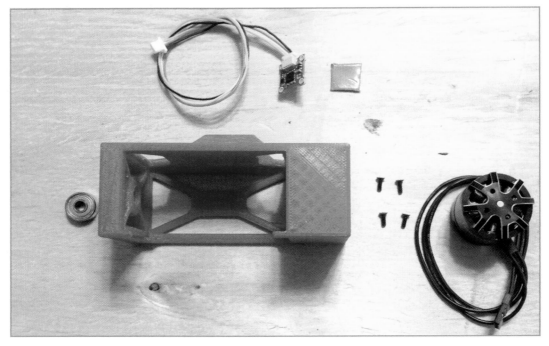

Figure 3-2 *Tilt Body assembly parts*

To assemble the Tilt Body, you'll need the parts pictured in Figure 3-2. First, you'll need to print out *Tilt-Body.stl*.

You'll also need one brushless motor, a 624zz bearing, four M2 x 6mm flat head socket cap screws, the IMU board (it's the small red board that came with the SimpleBGC gimbal controller kit) and a 1/2" square piece of double-sided tape.

MOUNT THE IMU BOARD

To mount the IMU board, cut a 1/2″ square piece of double-sided tape and place it on the underside of the board. Then press the IMU tape side down on the Tilt Body as shown in Figure 3-3.

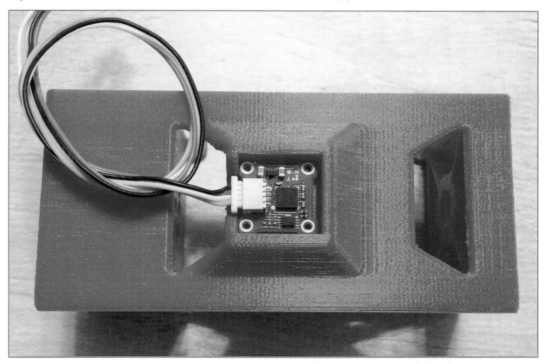

Figure 3-3 *IMU mounted in Tilt Body*

INSERT THE MOTOR

Insert the motor into the side of the Tilt Body as in Figure 3-4.

Ensure that the holes on the top of the motor align with the holes in the Tilt Body. Once aligned, secure the motor in place with four M2 x 6mm flat head screws.

Your screws should sit flush as in Figure 3-5 so they do not scratch your camera.

Figure 3-4 *Motor inserted into Tilt Body*

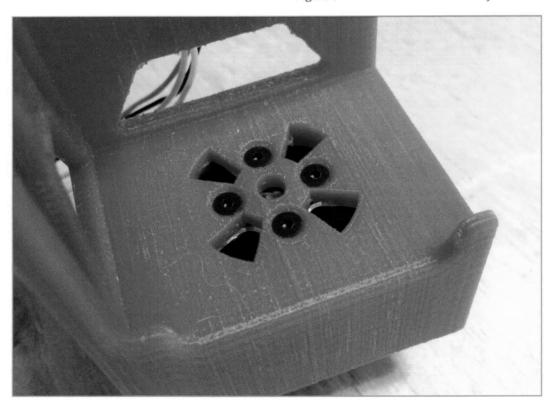

Figure 3-5 *Flush screws inserted*

SEAT THE BEARING

Next, seat the 624zz bearing into the right side of the Tilt Body (Figure 3-6).

Figure 3-6 *Seating the 624zz bearing*

Make sure the bearing is pressed all the way in; it should sit flush when seated.

The Tilt Body is now assembled (Figure 3-7), and will be used in the next step.

Figure 3-7 *Assembled Tilt Body*

ROLL BODY ASSEMBLY

Now you'll connect the Tilt Body to the Roll Body.

The Roll Body allows the camera gimbal to compensate for movement along the y-axis, or left and right from the camera's perspective.

First, you'll need to print out the *Roll-Body.stl* file.

The other nonprinted parts you'll need are the assembled Tilt Body, four M3 x 6mm socket head cap screws, and one M4 x 2mm socket cap screw (Figure 3-8).

ROUTE THE MOTOR WIRES

First, you'll need to feed the motor wires from the Tilt Body through the cavity in the Roll Body as shown in Figure 3-9.

When putting the two assemblies together, be sure to keep tension on the motor wires so that you don't crimp the wires by mistake.

Once together, you should no longer be able to see the motor from the Tilt Body (Figure 3-10).

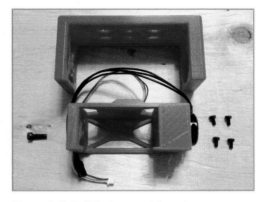

Figure 3-8 *Roll Body assembly parts*

Figure 3-9 *Feeding the motor wires*

Figure 3-10 *Tilt Body and Roll Body aligned*

SECURE THE MOTOR MOUNT

Ensure that the mounting holes on the bottom of the motor are aligned with the holes in the Roll Body, then secure the motor using four M3 x 6mm socket head cap screws (Figure 3-11).

The screw heads should be flush with the plastic.

Figure 3-11 *Adding the mounting screws*

SECURE THE TILT/ROLL BODY ASSEMBLY

Now we will secure the whole assembly with one M4 x 12mm socket head cap screw as shown in Figure 3-12. When you are threading in the screw, it is normal to feel some resistance (as if the hole was printed too small).

The hole is meant to be self-tapping, meaning once the screw is in, it will not come out without you manually removing it. The screw will be secure and hold tightly in place without extra hardware.

The screw will sit flush and should be sitting in the middle of the bearing that we installed (see Figure 3-6) into the Tilt Body earlier in the build (Figure 3-13).

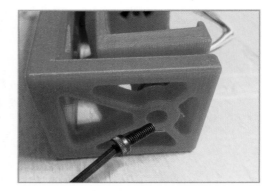

Figure 3-12 *Screw secures the two bodies*

The Tilt Body is now securely attached to the Roll Body, and you should be able to freely move the Tilt Body without any trouble as seen in Figure 3-14.

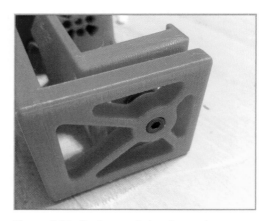

Figure 3-13 *Flush screw in bearing center*

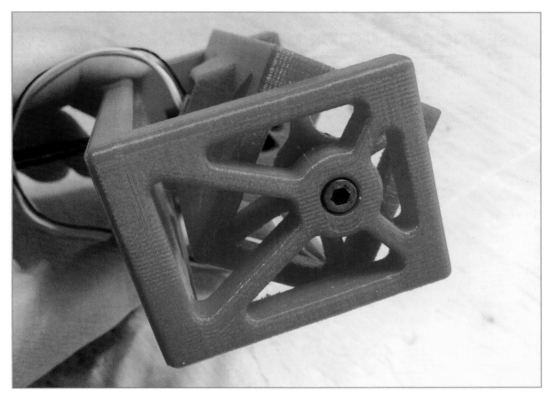

Figure 3-14 *Roll Body and Tilt Body move freely*

BASE AND ROLL MOTOR MOUNT ASSEMBLY

The base is the foundation of the camera gimbal. The Roll Motor Mount will be firmly secured to the Base, and will form the point of rotation for the system.

To add the Roll Motor Mount to the base, you'll first have to print out *Roll-Motor-Mount.stl* and *Base.stl*.

In addition, you'll need the brushless motor, and four M2 x 6mm and four M3 x 6mm socket head cap screws, as shown in Figure 3-15.

Insert the motor into the Motor Mount, and ensure that the holes line up as before. Then secure the motor to the Motor Mount with the M2 x 6mm screws (Figure 3-16).

The screw heads should sit flush with the surface of the Motor Mount. It will be mounted to the back of the Roll Body later and protruding screws will cause problems in the build.

Now you'll secure the Roll Motor Mount to the base. Ensure that the motor's wires are facing down as shown in Figure 3-17, and place the bottom of the motor into the socket of the base.

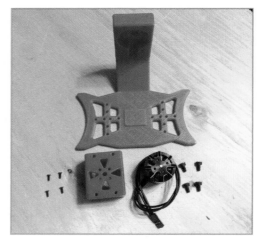

Figure 3-15 *Base and Roll Motor Mount parts*

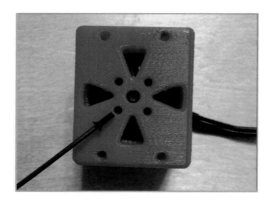

Figure 3-16 *Inserting the M2 x 6mm screws*

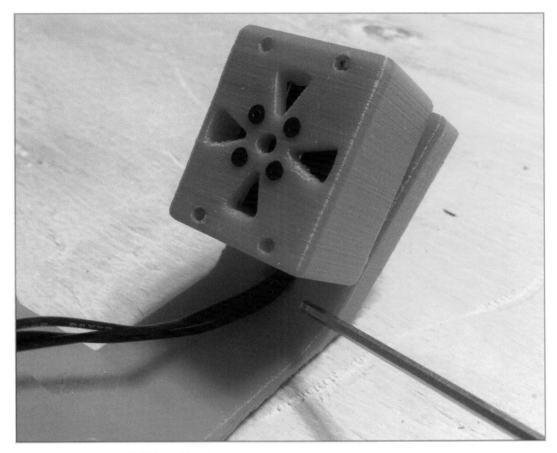

Figure 3-17 *Attaching Roll Motor Mount*

Using four M3 x 6mm socket head cap screws, attach the Roll Motor Mount (Figure 3-17 and Figure 3-18).

Figure 3-18 *The secured motor mount*

FINAL ASSEMBLY AND BALANCING

In this step, you'll put all the assemblies together and balance the Roll Body on the Roll Motor Mount.

At the end of the balancing process, the Roll Body should maintain its position after you've manually moved it with your fingers. If it continues to move after you remove hand contact, then it is not properly balanced.

Figure 3-19 *Parts to complete the gimbal*

 Balancing the gimbal is a very important step. Your camera gimbal will not work as intended unless it is balanced properly.

If the Roll Body stays where you positioned it, you have a well-balanced system.

In addition to the previously assembled sections, you'll also need four M3 x 16mm socket head cap screws and one GoPro Hero 2, as shown in Figure 3-19.

Insert two M3 x 16mm socket head cap screws into the top rail of the Roll Body (Figure 3-20), and two into the bottom rail.

As seen in the photos, each screw should be sticking out of the back of the Roll Body about 11mm (Figure 3-21).

Next, align the screws with the holes on the Motor Mount.

Figure 3-20 *Roll Body top rail screws*

Figure 3-21 *Aligning screws with Motor Mount*

Begin tightening all four screws, but don't tighten them completely. Just get them "tight enough" to secure the Roll Body to the Motor Mount (Figure 3-22).

Figure 3-22 *Motor Mount loosely attached*

 In order to balance the gimbal, the screws need to provide just enough play for the Roll Body to be able to slide back and forth. If you tighten them too much, you can't achieve a balanced gimbal.

Insert your GoPro. This will fully weight the gimbal and it will likely fall in one direction. Next, you'll manually balance the gimbal so it says put.

To balance the gimbal, slide the Roll Body back and forth on the screws that secure it to the Motor Mount until you find the point at which the body stays where you put it.

If you push the left side (Figure 3-23), you don't want it to continue moving when your finger stops, and you don't want it to fall back to the right either.

Figure 3-23 *Manually balancing the Roll Body*

The same goes for the right side. You want the assembly to stay in place after you manually move it with your finger. If it keeps moving, readjust the Roll Body's placement on the screws and try again until you find the correct positioning.

Once you have found the perfect spot, *very carefully* tighten down the screws. Congratulations, you now have a balanced gimbal!

Now you'll connect the control board to get the electronics set up and running.

ELECTRONICS SETUP

Figure 3-24 *Controller, unattached headers*

The SimpleBGC gimbal controller can be used with many different types of motors. For this project, you'll need to solder the kit's headers (shown unattached in Figure 3-24) to the board to make it compatable with the Turnigy HD 2212 brushless motors.

SOLDER THE HEADERS

Solder the headers to the underside of the board, as shown in Figure 3-25.

PLUG IN THE IMU BOARD

Next, plug the IMU board cable (attached to the Tilt Body, Figure 3-3) into the gimbal control board plug labeled "IMU." The connector is on the bottom of the board, as seen in Figure 3-25.

Add two pieces of double-sided tape to the bottom of the control board (Figure 3-26). This will keep it securely mounted to the base.

Figure 3-25 *Controller, soldered headers*

CONNECT THE MOTORS

Connect your motor wires to the control board. The ports are labeled on the bottom of the board, as shown in Figure 3-25.

In Figure 3-27, *Roll* is on the top left, and *Pitch* is on the bottom left. These connections are also printed on the underside of the board as *PIT* and *ROL*, visible in Figure 3-24 and Figure 3-25.

Figure 3-26 *Tape added to board's bottom*

Figure 3-27 *Motors connected to controller*

MOUNT THE BOARD

Mount the control board to the base, pressing down on the double-sided tape you added earlier.

Make sure to orient the motor connections toward the back as shown in Figure 3-28. This puts the micro USB port in the front, where it's easy to access, making tuning, programming, and occasionally updating firmware much easier.

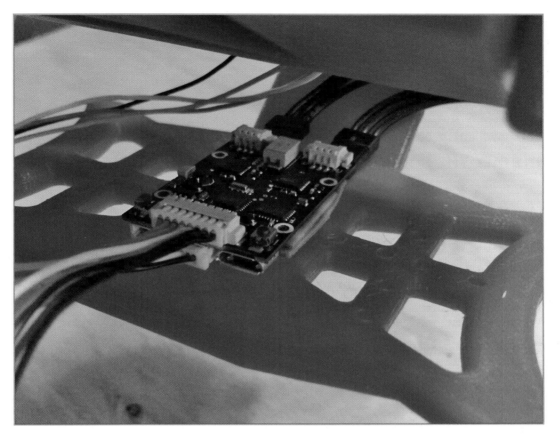

Figure 3-28 *Control board mounted on base*

Your gimbal is now wired and ready to be configured.

SOFTWARE CONFIGURATION

Now it's time to load new settings on the board that match both the gimbal and camera type to provide a stable picture.

The software screenshots shown in this project were taken in Windows, but both the gimbal configuration software and the necessary drivers are also available for Mac and Linux.

INSTALL THE SOFTWARE

Before you can configure your gimbal, you'll need to get the SimpleBGC 8-bit GUI (*http:// bit.ly/1FI2NFU*) from BaseCam Electronics; at the time of publication, the most recent version is v2.40b7 (*http://bit.ly/1KREzFh*).

INSTALL THE VCP DRIVERS

You'll also need the VCP drivers (*http://bit.ly/1Lk0CLh*) for your operating system, otherwise your computer won't recognize the control board.

On Windows, click the appropiate executable file for your OS, as shown in Figure 3-29. For Windows and Mac, the install wizards will guide you through the process.

OPEN THE SIMPLEBGC GUI

Unzip the *SimpleBGC_GUI_2_40b7.zip* file and launch the software. Steps are OS dependent, but on Windows just launch the *SimpleBGC_GUI.exe* file. Mac users will have to right-click the *SimpleBGC_GUI.jar* file and then accept that its from an "unidentified developer."

If you're on Linux, navigate to the folder and type **run.sh**.

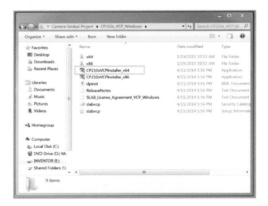

Figure 3-29 *VCP drivers folder*

If you have issues with the software or drivers, see the README.txt file included in the folder.

ADD THE BATTERY

Plug the battery into the last available connector on the gimbal control board.

Powering the board through your computer's micro USB connection doesn't provide enough power for the motors to function properly; they need 12V.

CONNECT AND BEGIN CALIBRATION

Attach a micro USB cable to the control board, then plug it into your computer to calibrate the IMU.

Once the software is open, connect to your control board by selecting the appropiate port and then click the "Connect" button as shown in Figure 3-30.

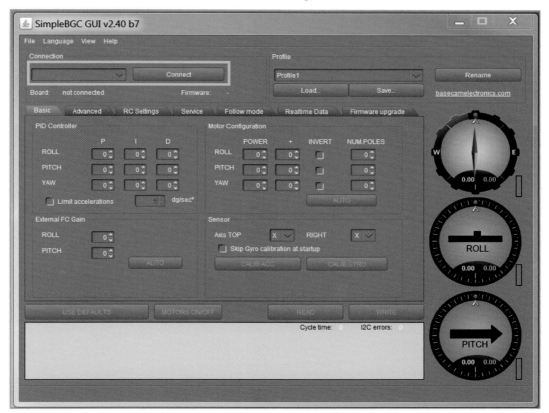

Figure 3-30 *Select the port*

On power up, the gimbal control board reads the sensor values and sets the readings it gets to zero. This means that if you power on your gimbal and your roll axis is tilted over to one side, the controller will assume that is level—or zero—even though it's not. It's important to roughly level the gimbal before powering up.

MODIFY THE CONTROLLER SETTINGS

Once connected, select the "Basic" tab as shown in Figure 3-31. Change the "PID Controller" and "Motor Configuration" settings to match those shown in the screenshot. Table 3-1 provides an easy-to-read version of the same information.

Table 3-1 *SimpleBGC software settings*

SOFTWARE SETTINGS			
PID Controller			
	P	I	D
ROLL	30	0.01	25
PITCH	17	0.03	12
Motor Configuration			
	POWER	+	NUM.POLES
ROLL	128	8	193
PITCH	84	8	12

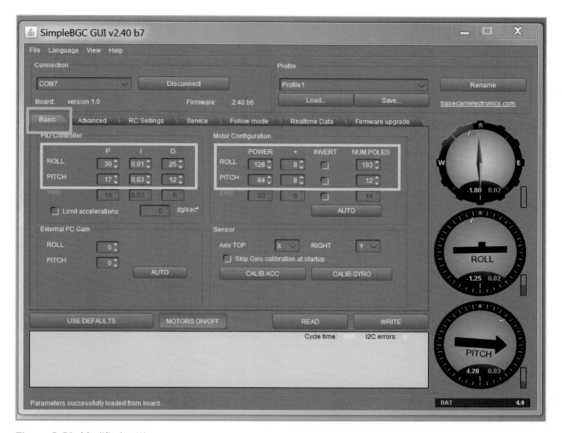

Figure 3-31 *Modified settings*

You can also use the "AUTO" tune button located in the middle right of the screen. It won't be perfect, but it will get you in the ballpark of how your gimbal should respond.

Once you have modified the settings, click the "WRITE" button (Figure 3-32) to save the new settings to the control board.

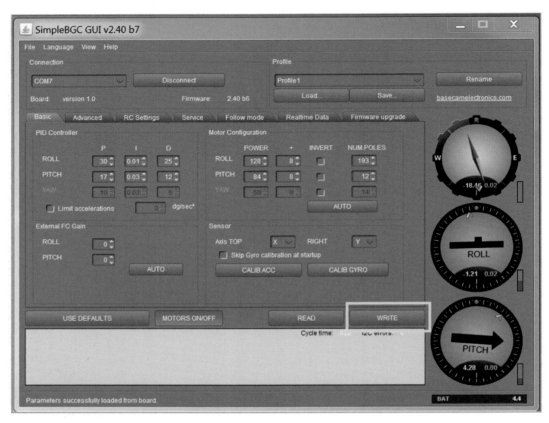

Figure 3-32 *Click "WRITE" to save*

 Make sure to click the "WRITE" button, or the new settings will not be saved!

Once finished, click the "Disconnect" button, highlighted in Figure 3-33, and unplug the USB cable that linked your control board to the computer.

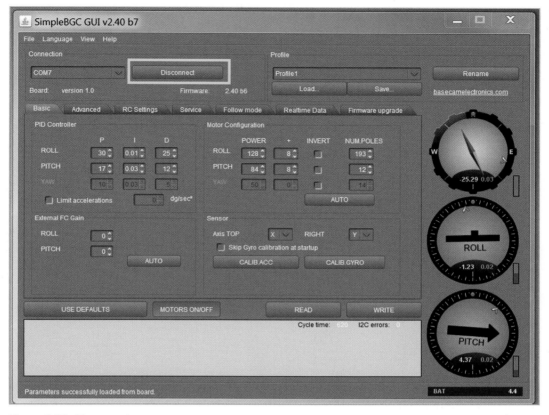

Figure 3-33 *Disconnect*

Congratulations, you are now ready to go capture stabilized, high-quality video!

BubbleBot

by John Baichtal

The BubbleBot is a robot that makes bubbles! Also known as a "bubble machine" to those content simply to buy them, the BubbleBot is cooler because you make it yourself!

This project uses a 3D printer to output a circular bubble wand that is rotated through a bath of sudsy solution, providing an unending stream of glistening bubbles at yard parties and other outdoor events.

COST
+ $70

PRINT TIME
+ 6 hours

BED SIZE
+ 2.7"x3"x0.4"

ASSEMBLY
+ A weekend

The BubbleBot even improves on most commercial bubble machines, giving you multiple customizable speed settings for three different default bubble styles. But the fun doesn't stop there! Thanks to desktop manfacturing, a few parts, and a little bit of code you can create limitless types of ephemeral orbs, from the miniscule to the voluminous.

PARTS, TOOLS, AND FILES

FILES TO DOWNLOAD

To complete the *Bubblebot* project, you'll need to download the fabrication files and code (*http://bit.ly/1VpUJM1*) from the *Make: 3D Printing Projects* site (*https://github.com/Make3DPrintingProjects*).

FABRICATION FILES

Quantity	Description	Filename
2	Star-shaped 3D-printed bubble wand	*wand_b.stl*
2	3 small holes 3D-printed bubble wand	*wand_c.stl*
2	Large hole 3D-printed bubble wand	*wand_d.stl*
1	Bubble wand hub	*hub.stl*
1	Chassis (ideally laser cut)	*bubblebot.svg*
1	Laser-cut acrylic box or a Tupperware tub	You'll create your own acrylic box later in the project

PARTS

Quantity	Part description	Part number
1	Squirrel cage fan	SparkFun 11270
1	Small reduction stepper motor: 3/16th shaft with a "D"-shaped profile to accomodate set screws	Adafruit 858
1	Axle coupler, 3/16" to 1/4"	SparkFun TK 12127
1	Axle, 1/4"x1" stainless d-shaft	SparkFun TK 12535
1	Hub, 1/4" set screw hub	SparkFun TK 12488
1	Adafruit half-sized Perma-Proto board	Adafruit 1609

PARTS		
1	TIP120 Darlington transistor	Adafruit 976
1	Male header pins with 0.1" spacing	Adafruit 392
1	Hall effect sensor	Adafruit 158
1	Rare-earth magnet	Adafruit 9
1	DC barrel power jack	SparkFun 119
1	5V power supply	Adafruit 276
1	2.2K and 10K resistors (SparkFun sells a great multipack)	SparkFun 10969
1	22-gauge solid hookup wire	Adafruit 1311
6	# 3 x 0.5" wood screws for bubble wand	
2	# 4 screws to attach the stepper	
2	# 4 x 1" screws, through front of hub	
3	# 4 x 2" screws to attach side plates to fan	
3	Zip ties (optional) can be used as a substitute for attaching the fan side plates	

TOOLS		
3D printer	Laser cutter access with a minimum bed size of 11" x 11"; see below for other options	7/32" x 4' x 8' birch underlayment, Home Depot PN 431178

FABRICATE THE CHASSIS

To build your own BubbleBot, you'll need to fabricate a wooden chassis, designed to be laser cut. The design could also be simplified for those handy with a saw.

A stepper motor bolted to the frame turns the bubble wand disk. You'll also need a special fan called a squirrel cage fan. Finally, you'll add the electronics, which consist of an Arduino, a transistor, and a magnet detector called a Hall effect sensor. Let's do it!

The unassembled laser-cut chassis parts are shown in Figure 4-1.

Once you have your parts, follow these steps to build the BubbleBot:

Figure 4-1 *Laser-cut the chassis or build something similar*

1. Find a laser-cutting service in your area or locate one online with reasonable shipping rates.

2. Use your carpentry skills to make a wooden box the approximate size of the enclosure, using the laser design as a pattern.

3. Use a building set like Actobotics (sold at SparkFun.com), Makeblock (makeblock.cc), or VEX (vexrobotics.com) to build a similar chassis.

4. Create an enclosure that your 3D printer can output. You'll have to extrude the SVG file and it's likely to take forever to print and use a ton of filament—but who else can say they printed their own bubble machine?

ASSEMBLE THE CHASSIS

Glue the sides to the back, bottom, and front. It should look like Figure 4-2.

Figure 4-2 *Assemble and glue together the chassis*

PAINT THE CHASSIS

Assuming you made your chassis out of wood, I suggest you spray a few coats of glossy paint on it, like I did in Figure 4-3.

Figure 4-3 *Spray-paint the chassis to help it resist moisture*

ASSEMBLE THE FAN MOUNT

Use the parts you lasered out with the chassis to make the fan mount. Attach the side plates to the fan with zip ties, as shown in Figure 4-4. A #4 x 2" screw would likely be better, but I didn't have any and zip ties worked perfectly well!

Figure 4-4 *The fan is attached to the mount with zip ties*

ATTACH FAN TO CHASSIS

Attach the fan assembly to the chassis, as shown in Figure 4-5. The wooden assembly glues into the chassis, and the fan's wires can be left to dangle for now.

Figure 4-5 *The fan assembly gets connected to the chassis*

ATTACH THE STEPPER

Using two #4 screws, attach the stepper using the mounting holes already lasered into the chassis. It should look like Figure 4-6. I made that hole way too big; my final laser files have a hole only a little bigger than the axle coupler.

Figure 4-6 *Two #4 screws attach the stepper*

ATTACH THE SHAFT COUPLER TO THE AXLE

Attach the shaft coupler to the axle. Secure the set screw to make sure the two parts don't move around.

Figure 4-7 shows the axle, hub, and coupler connected to the stepper just to illustrate how they come together; you'll want to wait until the wand is attached to the hub before you attach them to the motor.

Figure 4-7 *Axle, hub, and coupler*

MAKE THE BUBBLE SOLUTION RESERVOIR

Assemble the reservoir out of laser-cut acrylic, glued together with acetate. It should slide right into the box formed by the front and middle panels.

I'm going to leave it to you to design the reservoir, just to nudge you!

Go to makercase.com and enter the dimensions of the reservoir. My dimensions were 9" wide by 4.75" tall and 4.25" deep. Set the material thickness to 1/4" and keep the edge joints at flat (Figure 4-8).

The site will generate a file ready to laser. It will also include a top, but this can be deleted out of the file before you print. If you don't have access to a laser, I suggest building the BubbleBot around an existing receptacle like a Tupperware tub.

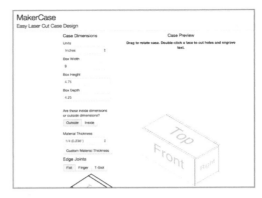

Figure 4-8 *Create an acrylic box*

PRINT THE BUBBLE WAND

Due to the limitations of my printer's bed size, as well as the possibility of being able to swap in and out different hole patterns, I printed the wand as a series of six triangular wands (Figure 4-9). You'll assemble them together with the help of a hub.

Figure 4-9 *Printing the bubble wand*

ASSEMBLE THE BUBBLE WAND

Thread #4 x 1″ screws through the front of the hub (see Figure 4-10) and through the corresponding mounting hole in the bubble wand parts.

Figure 4-10 *Assembled bubble wand*

ATTACH THE BUBBLE WAND

Secure the hub to the 3D-printed bubble wand using #3 x 0.5″ wood screws as shown in Figure 4-11. Then attach it to the axle, securing it by tightening the set screw.

Figure 4-11 *Attach the bubble wand to the motor's shaft*

The wood screws may not be the best solution for attaching the wand; the four mounting holes don't match up well with the six wands. I figured it would be easiest just to let you place the mounting holes yourself.

Next, attach the magnet to the wand using hot glue. For best effect, you'll want to make sure it comes within 0.25" of the Hall effect sensor, once that is installed.

SOLDER THE HEADERS

Let's begin wiring up the electronics.

Insert 15 male header pins into the digital pins on the Arduino, then put the protoboard (in the location shown marked in black strip across the pins) on the tops of the headers (Figure 4-12). Solder them into place.

 The reason you put the headers into the Arduino first is to keep them straight and even when you lay the protoboard on top.

Add the five male header pins for the stepper to column A, rows 6–10 on the protoboard as shown in Figure 4-13. This time, insert the pin short ends into the *top* of the protoboard and solder in from underneath.

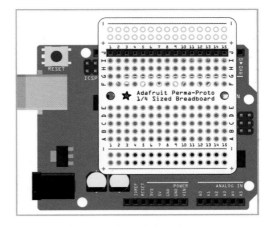

Figure 4-12 *Insert the male header pins*

SOLDER THE STEPPER WIRES

Solder four short wires from the first four stepper pins to the correct digital pins on the Arduino, shown in Figure 4-14.

The stepper has five wires; orange connects to digital 11 on the Arduino, yellow to 10, pink to 9, and blue to 8. The red wire doesn't connect to anything (see Table 4-1).

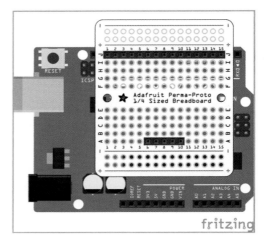

Figure 4-13 *Add pins for stepper*

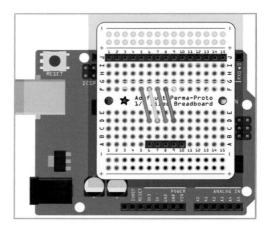

Figure 4-14 *Stepper pins to Arduino*

Table 4-1 *Stepper connections*

STEPPER TO ARDUINO		
Orange wire	→	PIN 11
Yellow wire	→	PIN 10
Pink wire	→	PIN 9
Blue wire	→	PIN 8
Red wire	→	not connected

ADD POWER CONNECTIONS

Attach the DC power plug. Its power pin (on the end) connects to the power bus of the board, while the ground pin (on the side) connects to the ground bus of the board.

While you're at it, run a wire from the power bus to the VIN pin on the Arduino. Figure 4-15 shows how it should look.

Next, solder in the transistor. This part consists of three pins, the base (control) pin, the collector (middle), and the emitter (right).

The base is the leftmost pin as you look at the part's front; the part is inverted in Figure 4-16.

Connect the base to Arduino PIN 13 via a 2.2K resistor. You may be able to simply use the resistor as a jumper, as shown in the figure.

The transistor emitter connects to ground, shown with a cyan wire. Leave the collector pin alone for now.

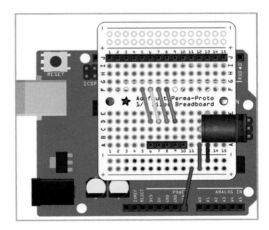

Figure 4-15 *Solder in DC power plug*

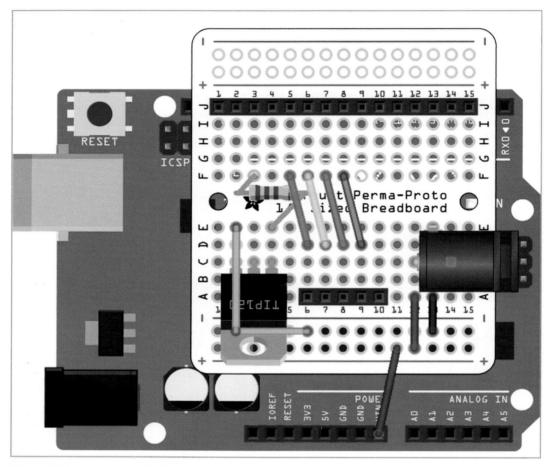

Figure 4-16 *Solder in the TIP120 transistor*

CONNECT THE SENSOR

Connect the Hall effect sensor. This component has three leads. The leftmost is 5V, the middle is GND, and the righthand pin is Data. Connect GND to the GND bus on the protoboard, shown with a blue-and-white striped wire in Figure 4-17.

The last two wires are tricky. They both connect to 5V, but only the Data *wire uses a 10K resistor to do so.*

I used an unpopulated row on the protoboard, row 11. I connected the row to the power bus, then connected the row to the sensor's 5V pin, shown with a red-and-white wire in the figure.

The sensor's Data pin connects to Arduino PIN 2, which is row 15 on the protoboard. From this same row, you need to connect to row 11 using a 10K resistor, as shown in Figure 4-17.

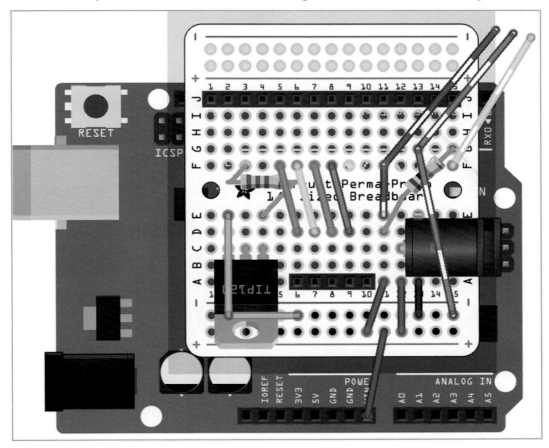

Figure 4-17 *Attach the Hall effect sensor*

ADD THE FAN

Lastly, let's add the fan.

 In Figure 4-18 the fan is shown as a DC motor. In the illustration, the motor's leads are depicted as green and yellow, but the actual fan should have red (positive) and black (negative) wires.

Connect the red wire to the power bus of the board, while the black wire connects to the collector (middle) pin of the TIP120 transistor.

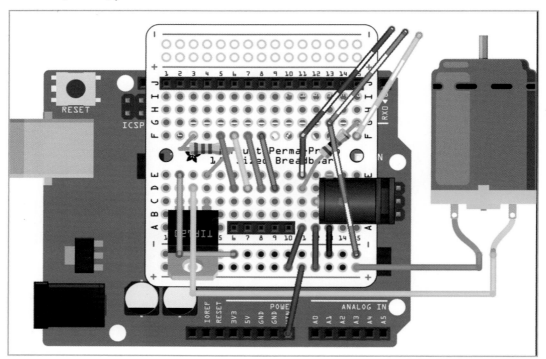

Figure 4-18 *Solder in the fan*

DOWNLOAD AND CUSTOMIZE THE CODE

Download the BubbleBot code (*https://github.com/Make3DPrintingProjects/BubbleBot*) from the Make: 3D Printing Projects book site.

Next, you'll program the BubbleBot. The sketch begins by calibrating the BubbleBot, which means rotating the wand until it detects the magnet with its Hall effect sensor.

Once the bot detects the magnet, it establishes that as `wandStep 1`, with a fan duration and speed set to that hole pattern.

If you want the three little holes wand to be blown quickly, dial in a quick burst. Conversely, for the big hole wand you may want a sustained, slow breeze.

Play with the settings in the Arduino sketch to get your wands to blow the type of bubbles you like best.

DDriver Rechargeable Screwdriver

by Rick Winscot

I got this great light-duty battery-operated screwdriver for my birthday last year. Sadly, after only a few days its little gears ground to a halt.

"Everyone finds a *dud* here and there," I thought. So I bought another—which lasted all of 10 minutes before it too stopped working. Rather than exercise my "right to replacement" via warranty, I decided to see what made these little tools tick. After all, warranties were meant to be voided.

The ugly truth was that both of my screwdrivers were beyond repair—and given the substandard quality, purchasing another didn't make sense. So I decided to disassemble the broken units, then use what I learned to build a new screwdriver from scratch.

COST
+ $35

PRINT TIME
+ 6 hours

BED SIZE
+ 6"x6"x6"

ASSEMBLY
+ 30 minutes

PARTS, TOOLS, AND FILES

FILES TO DOWNLOAD

To complete the *DDriver Rechargeable Screwdriver* project, you'll need to download the fabrication files and code (*http://bit.ly/1MEG25y*) from the *Make: 3D Printing Projects* site (*https://github.com/Make3DPrintingProjects*).

PARTS

Description	Part number
Adafruit Mini LiPo with Mini-B USB Jack	Adafruit 1905
USB cable A/Mini-B	Adafruit 260
Adafruit lithium ion cylindrical battery 3.7V 2200mAh	Adafruit 1781
Adafruit hookup wire 22AWG	Adafruit 1311
Adafruit heat shrink pack	Adafruit 344
Pololu Micro Metal Gearmotor HP	Pololu 994
DPDT rocker switch	Digi-Key CH941-ND
.47uF capacitor	Digi-Key 399-4309-ND
10 x 15 x 4 rubber sealed ball bearings (10 pack)	Traxxas TRA5119
M3 x 8mm stainless steel socket cap screw (100 pack)	McMaster-Carr 92125A128
M3 x 4mm steel set screw	McMaster-Carr 91217a055
Screwdriver set	Adafruit 822

TOOLS & MATERIALS	
3D printer	Soldering iron
Solder	Flux
Flush diagonal cutters	Digital calipers
Silver PLA filament	Gorilla Super Glue (gel)
Black PLA filament	220 grit sandpaper
Optional: Torch or heat gun for melting PETT filament	*Optional:* Taulman clear T-glase PETT filament

The interactive DDriver Tinkercad model (*http://bit.ly/tinkercadmodel*) (Figure 5-1) is avaliable online.

3D PRINTABLE FILES	
Quantity	**Filename**
1	*chuck.stl*
1	*handle.stl*
1	*handle_base_cap.stl*
1	*handle_base.stl*
1	*handle_top.stl*
1	*motor_brackets.stl*
1	*motor_cover.stl*

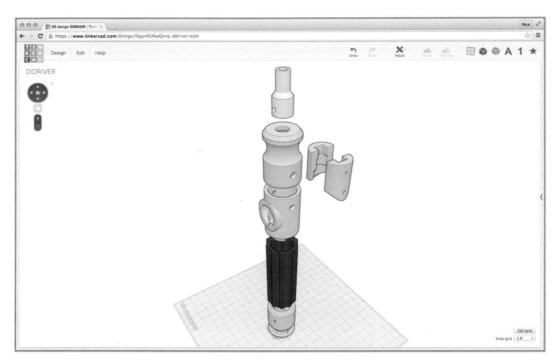

Figure 5-1 *Tinkercad DDriver model*

SCREWDRIVER TEARDOWN: CAN I FIX IT?

This is what I found inside that nifty little screwdriver: a motor, switch, and a bit of wire, about two dollars in parts as seen in Figure 5-2.

"Wow," I thought, "that's it?"

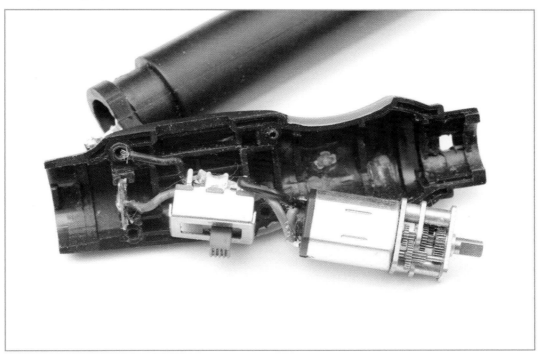

Figure 5-2 *Screwdriver innards*

EXPLORING THE INNARDS

Naively, I thought that I might be able to free the seized gears and get the screwdriver running again.

Unfortunately, this was impossible, as there were *"No Serviceable Parts"* (see Figure 5-3).

One item that I found interesting was the use of a double-pole, double-throw (DPDT) switch (Figure 5-4) to change polarity going to the motor: top points for cost-cutting creativity!

Manufacturers will go to great lengths to shave a penny or two off production costs and quality can quickly degrade.

A further inspection of the motor (Figure 5-5) shows that a 0.1uF capacitor was used to filter the commutator brush noise (produced as the motor shaft rotates). Given the load potential, I expected to see a capacitor with a value about three times greater. Compromises add up!

CAN I JUST REPLACE THE MOTOR?

Pololu has a wide variety of mini gearmotors that look perfect for the job. They have a matrix (*http://bit.ly/pololumatrix*) of options that will satisfy nearly any application.

I needed both speed and more torque than the original motor provided. Configurations typically favor either speed or torque, so you may have to strike a compromise. I decided to use the Pololu #994 Micro Metal Gearmotor HP.

Unfortunately, the motor in the broken screwdriver has a unique/nonstandard shaft length (Figure 5-6). Sigh.

Figure 5-3 *Damaged gear assembly*

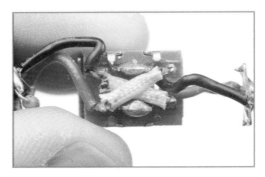

Figure 5-4 *DPDT switch*

Figure 5-5 *0.1uF ceramic capacitor*

Figure 5-6 *Shaft length comparison*

WE CAN REBUILD IT!

We have the technology. We can make it better than it was before. Faster, stronger, more durable—and 3D printed!

My goals for the redesign were to:

- Improve maintenance and serviceability
- Reduce shaft *wobble*
- Ditch the switch
- Improve torque
- Increase running time

I started out by modeling the screwdriver body in Tinkercad. Fire up your WebGL-compliant browser to take a look at the initial draft of DDriver (*http://bit.ly/tinkercadmodel*).

My initial plan was to produce a custom PCB built around the MCP73831 Li-Ion/Li-Polymer Charge Management Controller by Microchip. Instead, so others could easily replicate this project, I decided to use Adafruit's Mini LiPo w/ Mini USB breakout, as shown in Figure 5-7.

SOLDER WIRES TO THE BREAKOUT

Solder a length of red (positive) and black (negative) wire to the battery and ground pins, respectively. You'll run these up and through the battery compartment to the DPDT switch above. Cut your wire a little long; trimming is easy—adding length isn't!

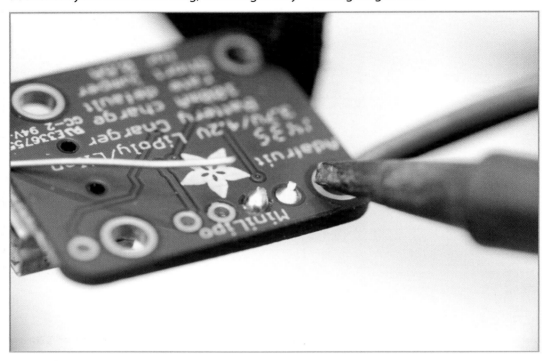

Figure 5-7 *Soldering battery leads*

Before you turn your soldering iron off, take a look at the breakout charge-rate jumper pad, as seen in Figure 5-8 opposite the newly soldered wires and marked with an upside-down "500mA." When open (default), it will charge your battery at 100mA. If it is closed with a solder bridge, it charges at 500mA.

100MA VERSUS 500MA?

If you plan on charging the screwdriver from a standard USB port, the default 100mA option is probably best and you don't need to take any additional steps.

Higher amperage will charge your DDriver faster. If you have a wall USB charger that can provide 1000mA or more, slap some solder on that pad to enable 500mA charging!

Figure 5-8 *Jumper adjusts charge rate*

PREP THE PLASTIC PARTS

Minimize gaps between mating parts by sanding each with a small square of 220 grit sandpaper (Figure 5-9).

 Here's a trick: don't blow the dust away. When superglue hits the dust it creates a slurry that fills gaps and dramatically improves plastic-to-plastic adhesion.

Figure 5-9 *220 grit sandpaper*

LIGHT-PIPE UPGRADE (OPTIONAL)

You can add a bit of light-pipe to make the charge controller indicators in *handle_base.stl* more visible (Figure 5-10). Taulman clear T-glase (PETT) works great.

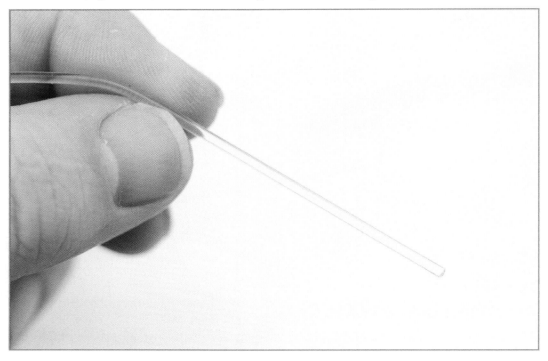

Figure 5-10 *Clear PETT light-pipe*

Hold a torch a few inches away until the filament begins to slump, put the torch down, grab the end of the filament, and pull it straight (gently) until it hardens.

Cover the filament with a bit of 2mm shrink tube on the inside of the housing to help isolate light from the indicator LEDs (Figure 5-11). A quick dash with the torch over the shrink tube in the housing will prevent the light-pipe from working its way out.

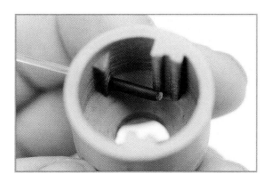

Figure 5-11 *2mm shrink tube*

Trim the filament to ~1mm above the housing (Figure 5-12) and quickly run a torch over the exposed ends (Figure 5-13) for polished perfection!

 Although Figure 5-12 and Figure 5-13 show the board inserted and trailing wires, I found it was much easier to add the shrink-tubed light-pipe first. You'll install those parts in the next step.

Figure 5-12 *Trimmed filament*

Figure 5-13 *Heat-polished filament*

ADD THE ELECTRONICS

Slide the Mini LiPo charger into the handle base, as shown in Figure 5-14. Make sure that the LEDs are aligned with the light-pipe.

Make sure the charger wires are sticking out through the top of the handle base; you can see them in Figure 5-13 and Figure 5-15.

Then glue the handle base cap (*handle_base_cap.stl*) into place.

Figure 5-14 *Mini USB charger installed*

On the opposite side of the handle base, thread in the battery wires (connector first) so you'll be able to plug them into the LiPo charger board later in the project.

DO NOT CONNECT THE BATTERY TO THE CHARGER NOW

For saftey reasons, ensure your battery isn't connected to the charging controller during soldering—and there's still a bit more soldering to do.

Next, swirl the extra battery wire into the handle base a la pigtail as shown in Figure 5-15. It's good to have extra; you need to borrow a little when it comes time to wire up the switch and motor.

Figure 5-15 *Pigtail wire as needed*

Slide the handle (*handle.stl*, which doubles as the battery housing) over the battery and charger wires, leaving the charger wire ends sticking out of the top of the handle.

Then screw the handle to the handle base slow and steady. Warm the screw a bit if it protests (Figure 5-16), but don't screw it down too tight, just snuggly. You'll need to open it up later to connect the battery.

Figure 5-16 *8mm socket cap screw*

You want to minimize *wobble* where the motor shaft meets the screwdriver; this reduces the potential for cross-threading and drilling oblong holes when using your DDriver.

Add one rubber sealed ball bearing to the screwdriver chuck (*chuck.stl*), shown in Figure 5-17.

When incorporating precise, tight-fitting parts, it can be tricky to get them to fit together properly. To ease the assembly process, and avoid cracking parts, warm the motor cover (*motor_cover.stl*) slightly with a torch to soften the plastic.

Then use the chuck to carefully seat the bearing flush with the surface of the freshly warmed motor cover as shown in Figure 5-18 and Figure 5-19.

Figure 5-17 *Screwdriver chuck with bearing*

Remove the motor cover and secure the set screw (Figure 5-20), but don't make it too tight at this point.

The final adjustment that optimizes the contact between the shaft adapter and bearing will happen later in the project.

Figure 5-18 *Using chuck for bearing install*

Figure 5-19 *Seated bearing*

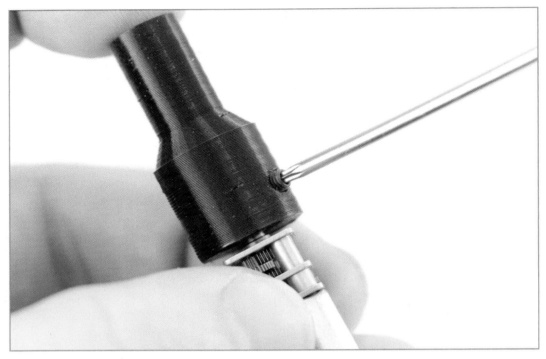

Figure 5-20 *M3 x 4mm set screw*

For safety, put some shrink tubing over the ends of the charger power and ground wires (Figure 5-21) sticking out of the handle.

Then route them through the button housing (*handle_top.stl*) as shown in Figure 5-22 and screw the handle into the button housing.

Figure 5-21 *Shrink tube on wire ends*

Solder a length of wire to each of the motor contacts with a .47uF ceramic capacitor between them as shown in Figure 5-23 and shrink-tube it!

Polarity isn't a concern since we will be driving the motor both forward and backward, so I used two blue wires instead of red and black ones. This also helps to color-code the wires later when you wire up the button.

Figure 5-22 *Side-ports prevent pinched wires*

Figure 5-23 *0.47uF capacitor across motor*

Sandwich the motor between the motor mount halves (*motor_brackets.stl*), then insert the mount into the motor cover (*motor_cover.stl*). Make sure to align screw holes visually as shown in Figure 5-24 (but don't add the screw yet!).

Remove the motor and give your set screw a final turn. Then reinsert the mount and screw in place (Figure 5-25).

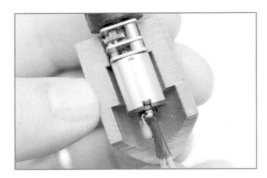

Figure 5-24 *Testing motor bracket fit*

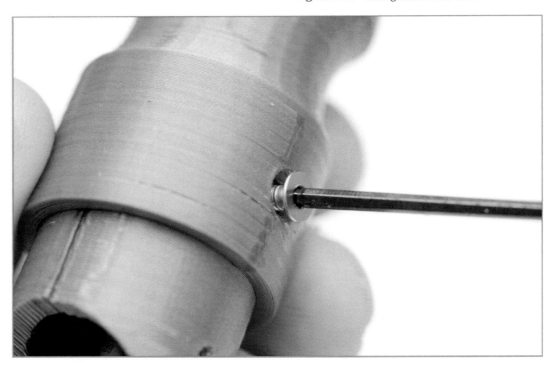

Figure 5-25 *Screwing the motor cover into place*

Pull the motor and charger wire "pigtails" through the large hole in the button housing and screw the motor bracket end into the button housing.

It's almost time to wire up the button (Figure 5-26).

Figure 5-26 *Power (red/black) and motor (blue)*

DPDT SWITCH BUTTONS!

I found a great DPDT button from Cherry we can use Figure 5-27; it's a big upgrade from the previous model.

The button's a little deep, but we can trim the switch connectors down to make it fit.

Add flux to your button connectors. Then you'll solder pigtails to your connections to keep what's behind your button a little more tidy (Figure 5-26).

Solder two short wire pigtails to the bottom button connectors and add shrink tube (Figure 5-28).

Figure 5-27 *Trimmed switch connectors*

Criss-cross the short wires and solder them and the leads from your motor to the top connectors as shown in Figure 5-29.

 Make sure not to melt or otherwise damage insulation on the wires as this could result in a dangerous short!

Finally, solder your power and ground to the center button connectors.

Push the button into the mounting hole (Figure 5-30) and reconnect the battery (Figure 5-31).

Figure 5-28 *Shrink tube protection*

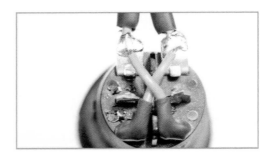

Figure 5-29 *Criss-cross polarity trick*

Figure 5-30 *Press-fit switch installation*

Figure 5-31 *Connect and charge battery*

Take one of the screwdriver bits and insert it into the chuck.

Give your DDriver a try!

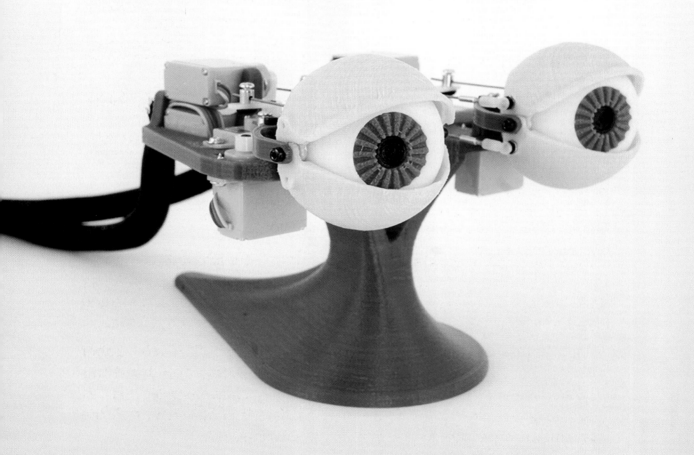

Animatronic Eyes

by Brian Roe

I was lucky enough to be a child in the 80s, when the incredible creatures of *The Dark Crystal*, *Gremlins*, *Harry and the Hendersons* and *Aliens* graced the movie screen—and stirred my imagination.

I was in the sixth grade when a career evaluation at my school pointed me towards a job as a Special Effects Technician. From this point forward I knew exactly what I wanted to be when I grew up! In high school I aspired to become an Animatronic Engineer, bringing fascinating characters to life through artistic robotics.

With a lot of hard work and a little bit of luck, I made it into the film industry and fulfilled my dream. Unfortunately for me, the industry was changing. CGI (Computer-Generated Imagery) was just catching on, eliminating the need to build expensive animatronic characters.

Although I no longer get to build crazy robots to stuff inside rubber monster skins for a living, I still love the challenge. I now build this stuff in my spare time. This Animatronic Eye mechanism project is just like one that I could have built for a creature in a film. It has all the movements necessary to create a very lifelike feel.

With a little practice, you'll be amazed how much emotion you can express with just a pair of eyes.

COST
+ $190

PRINT TIME
+ 8 hours

BED SIZE
+ 6"x6"x6"

ASSEMBLY
+ 8 hours

PARTS, TOOLS, AND FILES

FILES TO DOWNLOAD

To complete the *Animatronic Eyes* project, you'll need to download the fabrication files and code (*http://bit.ly/1iJfFB6*) from the *Make: 3D Printing Projects* site (*https://github.com/Make3DPrintingProjects*).

PARTS		
Quantity	**Part description**	**Part number**
1	2.4Ghz digital proportional 6 channel transmitter and receiver system	FlySky FS-T6
8	HK15168 coreless analog micro servo 8g	HobbyKing HK15168
8	30CM servo lead extention (JR) with hook (two packs of 5)	HobbyKing 015000008
1	Turnigy 3A UBEC w/ noise reduction	HobbyKing TR-UBEC
1	Arduino UNO R3	Adafruit 50
1	Adafruit 16-channel 12-bit PWM/servo shield - I2C interface	Adafruit 1411
1	9VDC 1000mA regulated switching power adapter	Adafruit 63
1	Premium female/female jumper wires - 20 x 6"	Adafruit 1950
1	Breakaway 0.1" 36-pin strip male header	Adafruit 392
4	Micro E/Z Links (one pack of 4)	Du-Bro 849
8	Mini E/Z Connectors (one pack of 12)	Du-Bro 915
4	Micro Ball Links for .032" wire (two packs of 2)	Du-Bro 928
1	.032" x 36" music wire	K&S Precision Metals 501
8	M2 x 4mm socket head cap screws (two packs of 4)	Du-Bro 2111

PARTS		
12	M2 x 6mm socket head cap screw (three packs of 4)	Du-Bro 2112
4	M2 x 10mm socket head cap screw (one pack of 4)	Du-Bro 2113
4	M2 x 12mm socket head cap screw (one pack of 4)	Du-Bro 2114

 If you use a different transmitter/receiver, it must have a switch to operate channel 5.

TOOLS & MATERIALS		
3D printer	Printer filament	Dremel rotary tool
1/16″ drill bit	Micro Phillips screwdriver	Small flat head screwdriver
Soldering iron	Solder	Needle-nose pliers
1.5mm allen wrench	Wire cutter/stripper	Music wire cutters
Cyanoacrylate (superglue)	Safety glasses	

3D PRINTABLE FILES			
Quantity	Filename	Description	Color shown
1	*Pupils.stl*	The eye's pupils	Black
1	*Iris.stl*	The iris	Blue
1	*Eyes.stl*	The eyeballs	White
1	*Eye_Lids.stl*	The eyelids	Flesh
1	*Mechanics.stl*	All the parts necessary to build the mechanical system	Gray
1	*Stand.stl*	The stand that supports the eye mechanism	Gray

3D PRINTABLE FILES

I have supplied the 3D models in six different files, allowing you to print the different parts in the PLA color of your choosing (see Figure 6-1).

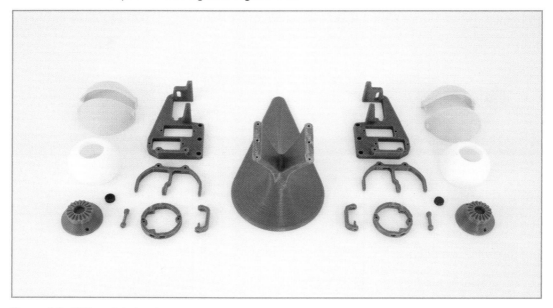

Figure 6-1 *All the printed parts*

 Each file contains both sets of parts shown, so although you need two of every part (except the stand), you only need to print each file one time.

These are very small parts; make sure your printer is well calibrated and in good working order.

None of the parts require support, except the eyelids, which have been modeled with the necessary support built into the design (see Figure 6-2).

After printing the eyelids, carefully break out the vertical supports with a pair of needle-nose pliers. Then get out your Dremel and use a round-shaped burr and a barrel-shaped burr to clean up the ridges left behind.

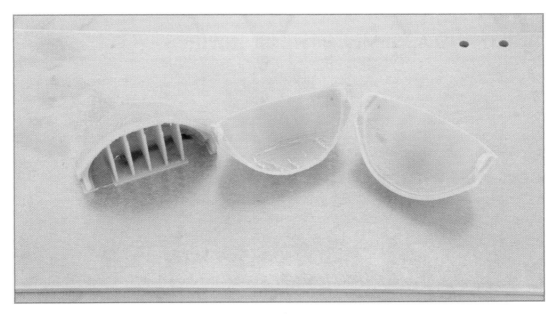

Figure 6-2 *Left: built-in support; right: support removed*

  Eyelid support cleanup is a very critical step; take your time and get it right. Any material left behind will interfere with the eye movement and the closing of the lids.

BUILD AND ASSEMBLE THE LINK WIRES

For this next section, you'll need the music wire, Micro E/Z Links, needle-nose pliers, music wire cutters, and cyanoacrylate.

You'll be using these materials to create four "L"-shaped links of two different lengths (42mm and 58mm). These wires connect the servo motors to the mechanical components to create the eye movements.

BEND WIRE WITH PLIERS

To create the first 42mm link, take the small pair of needle-nose pliers and bend the end of the music wire 5mm at a 90-degree angle, forming a "L" shape, as shown in Figure 6-3.

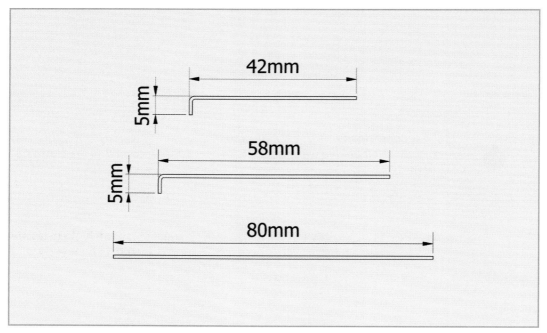

Figure 6-3 *Link wire lengths*

CUT WIRE TO LENGTH

Using the music wire cutters, cut the opposite end of the wire at 42mm. Repeat the process to create a second 42mm "L" link.

REPEAT FOR DIFFERENT LENGTHS

Using the same process, create two 58mm "L" links.

The "L" links will be used with the Micro E/Z Links to connect the servos and enable the up/down and left/right eye movements.

Cut "straight" 80mm Links

Next, trim four straight music wire pieces at 80mm (no "L" bend, see Figure 6-3). These will be used to create the linkages for the eyelids.

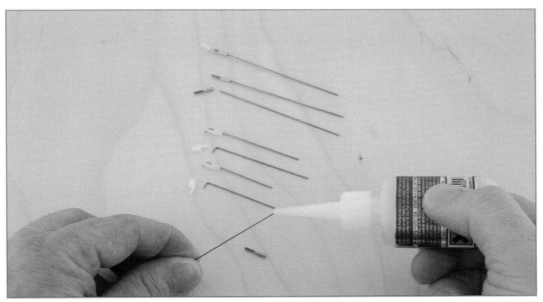

Figure 6-4 *Gluing the Micro Ball Links to wires*

GLUE THE STRAIGHT 80MM WIRES TO MICRO BALL LINKS

Remove the brass threaded adapters from the Micro Ball Links and attach them to the ends of the straight music wire with a small dab of cyanoacrylate as seen in Figure 6-4.

Screw the ball links approximately halfway onto the brass ends.

ASSEMBLE THE EYEBALL GIMBAL

Four of the parts that make up the eye gimbal are "mirrored" parts. They look very similar, but there are slight (and important) differences.

 To avoid confusion, I marked the files with indentations to differentiate the parts used in the left eye versus the right.

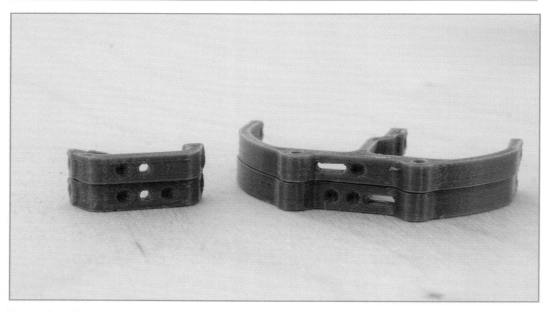

Figure 6-5 *The parts stacked on top (one divot) are for the right eye while those on the bottom (two divots) are for the left*

Parts to be used in the right eye will have a single divot printed into them while parts for the left eye will have two divots printed into them (Figure 6-5). All the other parts are interchangeable between left and right or it is very obvious where they are used.

CONNECT THE "L"-SHAPED LINKS TO THE DRIVE BARS

Thread the 42mm links into the small L/R drive pins and secure with the E/Z Links.

Insert the 58mm links into the up/down drive bars. Make sure to install them into the tab from the outside, as shown in Figure 6-6. After insertion, secure with the E/Z Links.

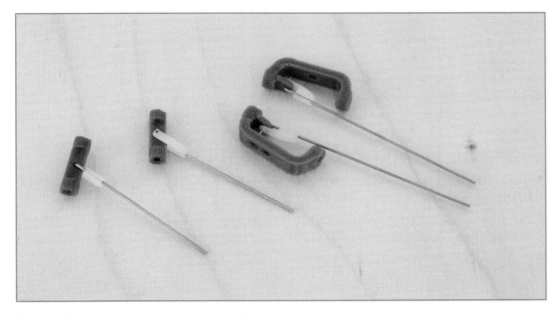

Figure 6-6 *Assembled L/R drive pins and up/down drive bars*

INSTALL THE BALL STUDS INTO THE EYELIDS

The lids have holes on both sides to accommodate left or right installation. Assemble them to create a mirrored pair as in Figure 6-7. Finish by snapping the ball links to the ball studs.

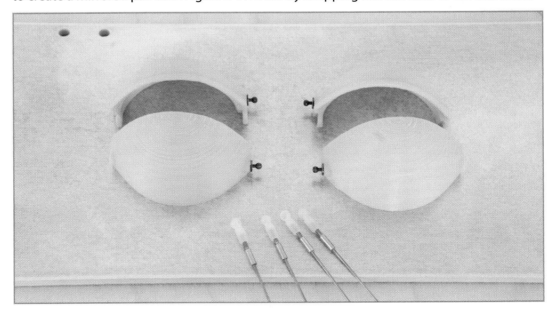

Figure 6-7 *Attach straight 80mm links*

Be careful not to crack the lids while attaching the ball links.

Notice that the upper eyelid has a small lip that allows it to hang over the lower lid. This gives the lids a visually interesting look and allows for a clean fit when they are in the closed position.

SCREW THE UP/DOWN DRIVE BAR TO THE E-BAR

Using a M2 x 6mm screw, connect the assembled up/down drive bar (Figure 6-6) to the E-bar. The link wire will connect to the top of the eye.

Make sure to place the single-divot right eye parts together and the double-divot left eye parts together. The parts shown in Figure 6-8 have the correct left and right orientations.

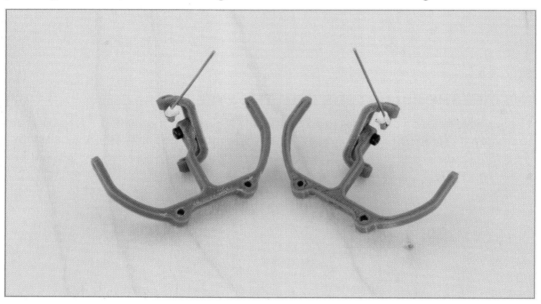

Figure 6-8 *Up/down drive connected to E-bar*

 When installing the screws be very careful not to overtighten them; they can strip easily. If you do happen to strip one of the holes during assembly, just reprint the part and try again.

The screw in Figure 6-8 is creating a pivot point for the up/down movement of the eye. Tighten it just enough to bring the parts together, but still allow for free movement. The parts should rotate freely but not have any slop between them.

INSTALL THE OUTER GIMBAL RINGS

Connect the outer gimbal rings to the up/down drive bars using two M2 x 4mm screws as seen in Figure 6-9.

Connect the L/R drive pin into the outer ring using two M2 x 4mm screws.

Be sure to run the link wire through the oval slot in the E-bar as you install the pin.

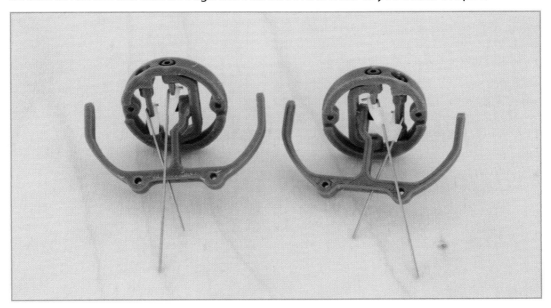

Figure 6-9 *Outer gimbal rings (with L/R drive pins installed) connected to the up/down drive bars*

 All four of these screws are pivot points, so tighten them to allow easy free movement without any slop.

Congratulations, you have just created a two-axis gimbal that fits inside an eyeball. Cool stuff!

ASSEMBLE THE EYEBALLS

Secure the pupil into the center of the iris with a small drop of glue (Figure 6-10).

The iris can then be inserted into the back of the eye as in Figure 6-11.

While looking through the small holes in the back of the iris, hold the assembly up to the light and rotate the iris inside the eye.

You will see the holes become brighter as they align with the holes in the eyeball. This is the position you need the parts in for the next step in the assembly.

Now insert each gimbal into the back of an eye and secure them with two M2 x 12mm screws (Figure 6-12).

Figure 6-10 *Pupils glued into the Iris*

Figure 6-11 *Iris inserted into eyeball*

Figure 6-12 *Gimbals inserted into eyeballs*

INSTALL THE SERVOS

Install four servos into each servo mount. Be sure to route the wires as shown in Figure 6-13.

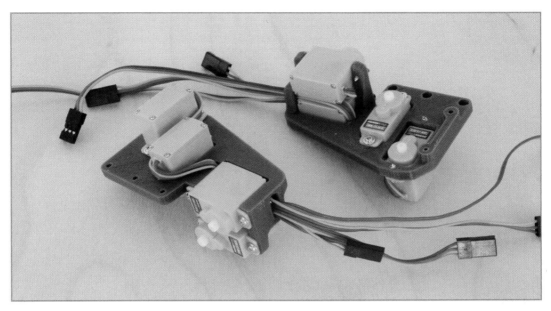

Figure 6-13 *Take note of the servo placement and wire routing, it's important!*

Later, you will use this wire positioning to determine the order the servos are plugged into the receiver. Use the small Phillips head screws supplied with each servo for mounting.

Be very careful not to strip out the heads of the screws. They are extremely soft metal and strip easily.

Use a screwdriver thats fits the head perfectly. If you do strip any screws remove them with needle-nose pliers and replace them with the M2 x 6mm screws we used previously.

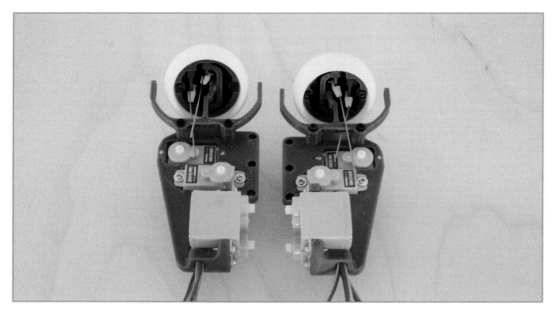

Figure 6-14 *Eyes attatched to servos*

Attach each eye to the servo mounts with two M2 x 10mm screws. Pay close attention to the divots on the E-bars and attach them to the correct servo mount as shown in Figure 6-14.

BUILD THE SERVO HORNS

Gather the straight single-sided servo horns from each servo pack. You'll use a 1/16" drill bit to open up the appropriate holes to fit the Mini E/Z Connectors.

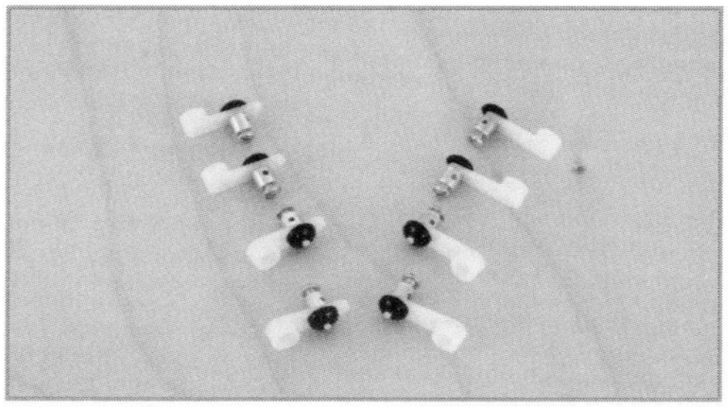

Drill out the third hole on four horns
On four of the horns, drill out the third hole from the center and install the connectors (as shown on the left side of Figure 6-15).

On the four remaining horns, drill the fifth hole
With the other four horns open up the fifth (last) hole from the center and install the connectors as shown on the right side of Figure 6-15.

Figure 6-15 *Drilled and connected servo horns*

SET UP THE ELECTRONICS

ASSEMBLE THE SERVO MOTOR SHIELD

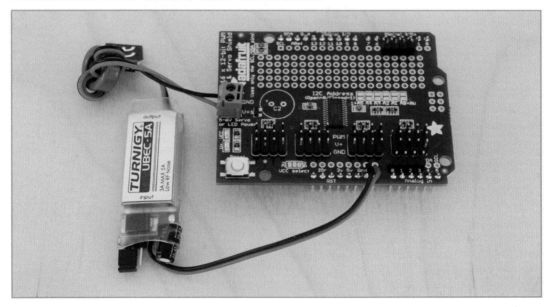

Figure 6-16 *Pins and power supply soldered to shield*

Assemble the servo shield kit using Adafruit's online tutorial (*http://bit.ly/servoshield*).

Remove four pins from the breakaway male header (purchased separately from the shield kit) and solder them to pins 2–5 on the digital I/O side of the board (see Figure 6-16).

Solder the input power wires of the Turnigy UBEC to the servo shield. The red input wire should be soldered to VIN and the black input wire should be soldered to the GND connection next to VIN as shown in Figure 6-16.

Trim the plug of the output wires and connect them to the power input connector on the servo shield. Be sure to get the polarity correct. Set the jumper on the input side of the UBEC to 5V.

With the servo shield soldering complete, stack it on top of the Arduino UNO. If you need help, see the Adafruit tutorial used to assemble the shield.

CONNECT THE SERVO SHIELD TO THE RECEIVER

Next, grab six female/female jumpers, one red, one black, and four random colors (these colors don't matter). You'll use the female/female jumpers to connect the servo shield to the receiver.

Plug the four colored jumper wires into the headers you add to pins 2–5 on the digital I/O (Figure 6-17).

Figure 6-17 *Motor shield stacked on Arduino*

Next, plug the black jumper into the GND pin servo position 15 and the red jumper into the V+ pin of position 15. See Figure 6-17 for reference.

Next, you'll connect the other end of the jumper wires to the receiver (Figure 6-18). A standard servo plug uses three pins. The brown or black wire of a servo is usually GND and located on one end of the plug. The center red wire is V+ or power. The wire on the other end (white or orange) is *signal*.

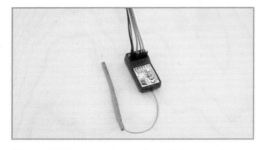

Figure 6-18 *Jumper wires installed on receiver*

Please reference your transmitter manual for the correct pin placement on your receiver. For the FS-T6, the signal pin is on the side closest to the antenna wire.

Plug the red wire into any of the center V+ pins on your receiver and the black wire into the matching GND- pin next to it.

Now connect the appropriate jumper wire from the digital out on the servo shield to the receiver signal pins as seen in Table 6-1.

Table 6-1 *Servo shield to receiver connections*

Servo shield pin		Receiver pin
Digi2	→	Chan1 Thro
Digi3	→	Chan2 Aile
Digi4	→	Chan3 Elev
Digi5	→	Chan5 Gear

When finished, the electronics assembly should look like Figure 6-19.

Plug the 9V DC power adapter into the power port on the Arduino to power up your receiver through the UBEC.

Follow the directions from your transmitter manual to bind the receiver to the transmitter.

Figure 6-19 *Final electronics assembly*

LOAD THE CODE

After the receiver has been bound you'll need to load the program (or sketch) onto the Arduino. You can download the sketch from this project's site (*https://github.com/ Make3DPrintingProjects/Animatronic-Eyes*).

In addition to the Arduino sketch, you'll need the Adafruit Servo Driver library (*http://bit.ly/ servlib*). Otherwise, you'll get error messages and won't be able to compile the program.

INSTALLING THE PWM SERVO DRIVER LIBRARY

You can easily add the Adafruit PWM Servo Driver library with the built-in Arduino Library Manager (Arduino Sketch menu→Include Library→Manage Libraries). Then search for "Adafruit PWM," install, and restart Arduino.

If you need step-by-step help with adding libraries, see Appendix A for more information.

FINAL CONNECTIONS

Remove the power adapter and USB cable from the Arduino.

Add all eight servo extensions to servo wires.

This is where the correct wire routing from the earlier step (see Figure 6-13) comes into play. Look at the length of the servo wires coming from the back of the eye mech. You will notice they are all a different length.

Start with the shortest wire on the right eye. Plug this servo into channel 1 on the servo shield.

Figure 6-20 *All wired up!*

Locate the pin labeled PWM on the shield. Connect PWM to the orange wire on the servo extension.

Continue to move down the channels with 2 being the next longest servo wire. Channel 3 is the next longest wire and channel 4 is the longest wire on the right eye.

For the left eye the shortest wire goes to channel 5 and so on like before with the longest wire on the left eye going to channel 8; see Table 6-2.

Table 6-2 *Eye servos: channel wire lengths*

Eye	Channel	Wire length
Right eye	Channel 1	Shortest wire
Right eye	Channel 2	Longer than 1
Right eye	Channel 3	Longer than 2
Right eye	Channel 4	Longest wire
Left eye	Channel 5	Shortest wire
Left eye	Channel 6	Longer than 5
Left eye	Channel 7	Longer than 6
Left eye	Channel 8	Longest wire

We are not using channel 0 on the shield, so do not start there.

INSTALLING AND CENTERING THE SERVO HORNS

With all eight servos connected you can reapply power to the Arduino from the 9V power adapter (Figure 6-20). After a few seconds Arduino will boot up and the servos will snap into position.

Be sure the right stick on your transmitter is in the center position and the left stick is in the lowest position. These are the neutral positions for installing the servo horns.

Figure 6-21 *Attaching the servo horns*

EYE SERVO HORN ADJUSTMENTS

For each eye, take two of the horns with the connector in the third hole and install them on the front two servos. Be sure to fish the link wires through the connectors as you install them.

Try to press the horns to the splines of the servo with the horn centered on the body of the servo. Sometimes the splines are not in the correct position; get them as close as you can. Figure 6-21 shows how the horns on my right eye ended up.

ADJUSTING THE OFFSET VALUES IN CODE

We will now use the offset values at the top of the Arduino sketch to bring the horns into the correct position. The sketch has all the values set to zero by default (Figure 6-22).

Let's start with changing the value of `#define RLRoffset` from 0 to 10. The nice thing about Arduino is that you can update the sketch while it's still running.

Go ahead and hit the Upload button at the top of the Arduino window. The sketch will be compiled and sent to the Arduino. You should see the servo jump to its new home position right away.

I like to move the stick on the radio a little to help the servo settle into its final position. Did the servo move the direction you need? If not try entering a -10 in the value and upload again. Maybe it moved the correct direction but not enough. Try a value of 20.

Spend some time playing with the values until you are happy with the horn positions. You can see my final result for these values in Figure 6-23.

```
#include <Adafruit_PWMServoDriver.h>
#include <Wire.h>

Adafruit_PWMServoDriver pwm = Adafruit_PWMServoDriver();

#define RLRoffset 0        // Right Eye Left-Right Offset
#define RUDoffset 0        // Right Eye Up-Down Offset
#define RTLoffset 0        // Right Eye Top Lid Offset
#define RBLoffset 0        // Right Eye Bottom Lid Offset

#define LLRoffset 0        // Left Eye Left-Right Offset
#define LUDoffset 0        // Left Eye Up-Down Offset
#define LTLoffset 0        // Left Eye Top Lid Offset
#define LBLoffset 0        // Left Eye Bottom Lid Offset
```

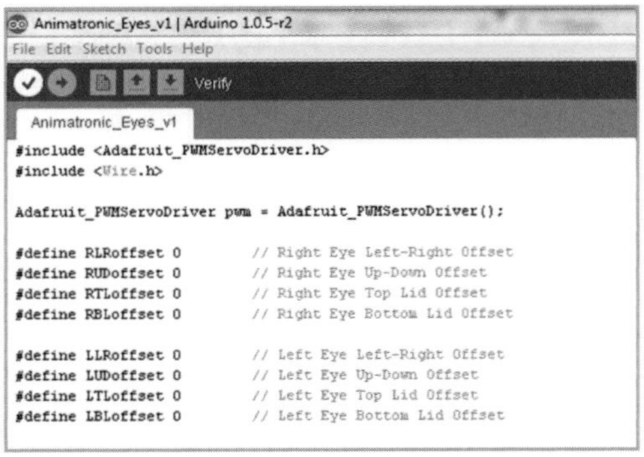

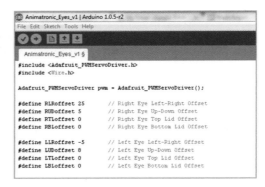

Figure 6-22 *Sketch values set to zero*

```
#include <Adafruit_PWMServoDriver.h>
#include <Wire.h>

Adafruit_PWMServoDriver pwm = Adafruit_PWMServoDriver();

#define RLRoffset 25       // Right Eye Left-Right Offset
#define RUDoffset 5        // Right Eye Up-Down Offset
#define RTLoffset 0        // Right Eye Top Lid Offset
#define RBLoffset 0        // Right Eye Bottom Lid Offset

#define LLRoffset -5       // Left Eye Left-Right Offset
#define LUDoffset 8        // Left Eye Up-Down Offset
#define LTLoffset 0        // Left Eye Top Lid Offset
#define LBLoffset 0        // Left Eye Bottom Lid Offset
```

Figure 6-23 *Servo offsets adjusted*

Figure 6-24 *Final servo horn positioning*

Those values gave me the final positioning of my servo horns, shown in Figure 6-24. Install the tiny screw that secures the horn to the servo (Figure 6-25).

Figure 6-25 *Tiny screws secure horn positions*

Now that the servos are correctly adjusted, we can lock off the link rods into the E/Z Connectors. Try to position the back of the eyeball square to the face of the servo bracket in both directions as in Figure 6-25.

Then tighten the flat head screw on the top of the connectors to lock the links in position. Do this for both the left and right eye.

When completed both eyes should be looking straight forward. Go ahead and give them a look around with the right stick!

EYELID SERVO HORN ADJUSTMENTS

You'll follow the same process with the lid servos. Install the remaining four horns on the lid servos and adjust the offsets accordingly. The image in Figure 6-26 shows the final position of the servo horns after adjustment.

Figure 6-26 *Final eyelid horn adjustment*

The horn closest to the eye should be slightly forward while the horn in the back should be perfectly vertical. Once you are happy with the positioning, install the horn screws.

INSTALL EYELIDS

You can now install the eyelids into the E-bars. Start by screwing an M2 x 6mm screw into each side arm. Allow the screw to protrude about 2 to 3mm through the arm.

Install the lids, making sure to insert the link wires into the E/Z Connectors on the lid servos (Figure 6-27).

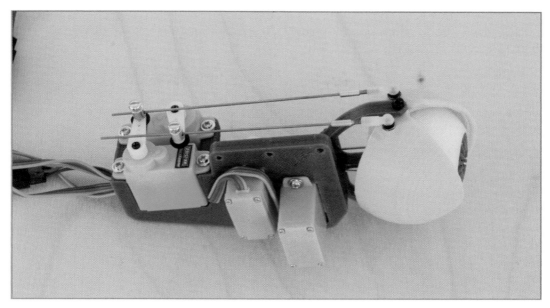

Figure 6-27 *Eyelids installed into the E-bars*

With the lids in place, finish tightening the side screws until the heads hit the side of the E-bar. These screws have now created a stud for the lids to pivot on.

STICK WITH IT!

Adjusting the servo horns, eyelids, and eyeball positioning can be a tricky, time-consuming process. Take the time to get it right. It will be well worth the effort in the final product!

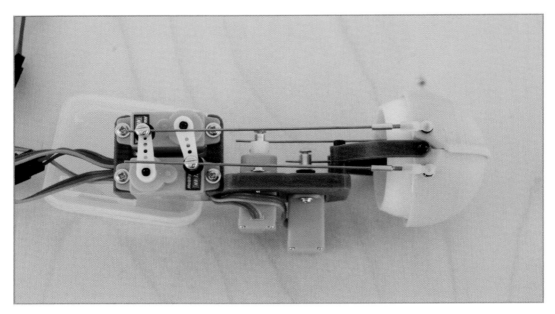

Figure 6-28 *Centering closed lids*

Pull the lids closed and position them centered in line with the E-bar arm (Figure 6-28). Tighten the flat heads on the E/Z Connectors to lock the lids in place (Figure 6-29). That's it!

Figure 6-29 *Tighten the screws*

TEST THE EYES

If you have adjusted all the servos correctly, you should be able to open the lids slowly with the left stick. Gently move the eyes around to all the positions.

Check that there is no binding and the servos are not straining. Cycle the eyelids open and closed a few times. The nice thing about having an Arduino in the loop is that you can create the custom channel mixing programs that allow the lids track the eyes as they move up and down.

Flipping the left switch on the transmitter will initiate a blink. You can do this at any time and the lids will return to their start position automatically after the blink.

The code is looking for a change in the state of the switch. Every time you flip the switch from front to back or back to front the eyes will blink.

The final step in the assembly is to secure each eye mechanism to the base using three M2 x 6mm screws as shown in Figure 6-30, Figure 6-31, and Figure 6-32.

Figure 6-30 *Secured to the base*

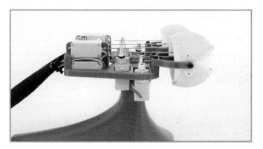

Figure 6-31 *Secured eyes, side view*

Figure 6-32 *Secured eyes, top view*

Figure 6-33 *Completed eye electronics*

ENJOY YOUR ACCOMPLISHMENT!

You have just created a very lifelike, radio-controlled, Arduino-powered, completely 3D printed set of animatronic eyes (Figure 6-33). Nice job!

I finished mine off with a little bit of wire management and some double-sided tape to hold the UBEC and receiver on top of the Arduino (Figure 6-34).

So what's next? I challenge you to create a larger much cooler looking base that can hold all the electronics along with a battery. This would allow wireless operation for when you would like to set the eyes on the mantle to impress your friends.

If you enjoyed the programming and electronics part of the project, I suggest you add an SD card reader to the mix so you can record and play back routines of your own.

You can also play with more servos—or try adding articulated eyebrows. You can really pump up the emotion with those!

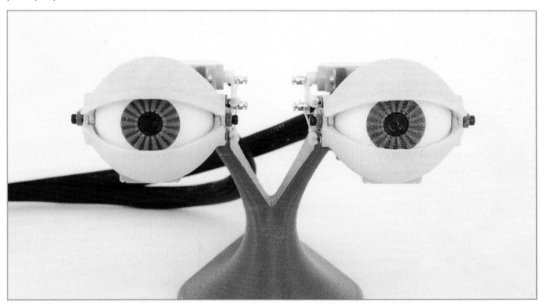

Figure 6-34 *Completed eyes*

Inverted Trike RC

by Steven Bolin

There are vast repositories of free 3D printable files, but nothing seems more exciting than a toy you would normally buy in some overpriced store.

The Inverted Trike mimics an expensive RC car, but keeps costs low by utilizing printed parts, commonly available hardware, and simple electronics. This quick and responsive RC car is easy to build and customize. Plus, if there's a collision, it's painless to print replacement parts.

COST
+ $100

PRINT TIME
+ 35 hours

BED SIZE
+ 6"x6"x6"

ASSEMBLY
+ 5 hours

The RC Inverted Trike was designed to fit within the Printrbot Simple's 6"x6" build volume, so that even those with smaller printers will still be able to print this car.

RC car enthusiasts of any experience level will enjoy the process of building and driving the Inverted Trike.

PARTS, TOOLS, AND FILES

FILES TO DOWNLOAD

To complete the *Inverted Trike RC* project, you'll need to download the fabrication files (*http://bit.ly/1VpV1m9*) from the *Make: 3D Printing Projects* site (*https://github.com/Make3DPrintingProjects*).

PARTS

Quantity	Part description	Part number
1	NTM Prop Drive Motor 750KV	HobbyKing NTM2830-750
1	Hitec HS-311 servo	RC Planet HRC31311S
1	Transmitter with receiver	HobbyKing GT-2b 9114000006
1	Turnigy nano-tech 2200mAh 3S 35-70C LiPo battery	HobbyKing N2200.3S.35
2	624zz bearings	McMaster 7804K103 (also check Amazon)
1	Turnigy plush 25-amp speed controller	HobbyKing TR_P25A
1	Set of four wheels (1:10 wheel and rim)	Amazon B00ID51PCG
1	Chuckable allen drivers (recommended)	A Main Hobbies PN PTK-8243

TOOLS & MATERIALS

3D Printer	PLA filament
NinjaFlex filament	2mm, 2.5mm, and 3mm allen wrenches
Needle-nose pliers	Micro Phillips screwdriver
Heat gun	Glue
1/4" heat shrink tubing	

HARDWARE	
Quantity	**Description**
3	M2.5 x 10mm
10	M3 x 8mm
12	M3 x 10mm
7	M3 x 16mm
2	M3 x 20mm
5	M3 x 25mm
2	M3 hex nuts
2	M4 x 12mm

3D PRINTABLE FILES			
Quantity	**Filename**	**Quantity**	**Filename**
3	Tires.stl	1	Back-Rim.stl
1	Back-Wheel-Mount.stl	2	Front-Rim.stl
2	Front-Rim-Cover.stl	2	Front-Bearing-Covers.stl
1	Body-Back.stl	1	Body-Front.stl
1	Bumper.stl	1	Frame.stl
1	Hub-Carrier-Left.stl	1	Hub-Carrier-Right.stl
2	Large-Shock.stl	1	Servo-Steering-Arm.stl
2	Small-Shock.stl	1	Stabilizer-Arm-Left.stl
1	Stabilizer-Arm-Right.stl	1	Steering-Hinge-Bottom.stl
1	Steering-Hinge-Top.stl	1	Steering-Hinge-Mount.stl
1	Steering-Hub-Left.stl	1	Steering-Hub-Right.stl
2	Steering-Linkage.stl	1	Wing-Left.stl (optional)

3D PRINTABLE FILES			
1	*Wing-Right.stl* (optional)		

 I recommend printing the shocks and stabilizer arms in NinjaFlex.

PRINTING TIPS AND PROFILES

Printing any functional part is a process of trial and error. These suggestions are based on my personal experiences using Cura to produce this project on my Printrbot Simple.

BASELINE SETTINGS

To save you time and effort, I've provided a print settings profile file, *Cura Profile.ini*, as a baseline (included in the project file download). As you attempt to print these parts on your machine, you may need to tweak the settings in order for your Inverted Trike to function to its highest potential, changing the infill and speed settings as needed.

My settings are also visible at a glance in the Cura screenshot, as seen in Figure 7-1.

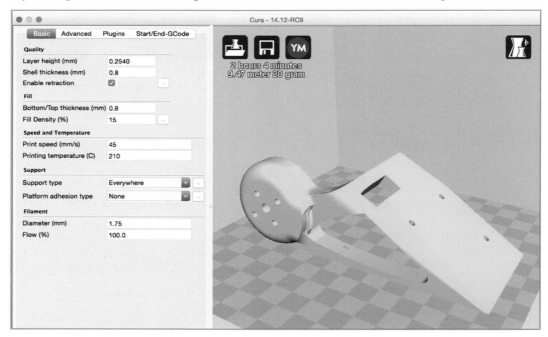

Figure 7-1 *Cura screenshot with settings*

CURA PRINT SETTINGS

Here are some additional factors that you'll want to take into account as you print the parts for the Inverted Trike RC:

Layer height

Based on how you want the car to look, you can adjust your layer height to give it a cleaner, smoother look by going to a smaller layer height, such as 0.1mm. If you'd like to avoid long print times, choose a layer height around 0.2mm to 0.3mm—especially for reprinting parts you may have broken in use.

Infill

Some parts require a little more strength and rigidity because of their size or function. The parts that need to be strong are the steering components, shocks, frame, and rear wheel mount. The bumper and wings need more layer height for strength and attractive top layer print surfaces. Using 15% infill gives you quality prints at faster speeds.

Shell thickness

I recommend a 0.8mm shell (at minimum) to strengthen parts and provide enough plastic for the screws to bite into.

Support

In order for some of the parts to fit and print well, the recommended Cura setting for support is "Everywhere" under the support type tab. Another setting worth checking is the "Structure Type" in the expert settings window. It gives you the "lines" support structure, which gives great results and is easy to remove.

Print orientation

It's pretty obvious how most parts should be oriented for printing. I suggest printing the front and back bodies upright in order for the layers to be consistent on your prints. It looks good as well as ensuring the desired fit.

When rotating your part, use the "Lay Flat" tool in order to get your base surface on the bed as flat as possible. This is very important when printing the back wheel mount because of the odd angles involved. Rotate the part to be as close to flat as you can get it, then hit "Lay Flat" and it will find its place. Although this part can print without support, you will guarantee your success by keeping the "Support Type" clicked on "Everywhere."

FRONT WHEEL ASSEMBLY

After printing all the wheel parts (*Tires.stl*, *Front-Rim.stl*, *Front-Rim-Cover.stl*, and *Front-Bearing-Covers.stl*), start by mounting your tires.

Whether you printed the tires with NinjaFlex, as shown in Figure 7-2, or you bought some off Amazon (Figure 7-3), mount the tires and glue them to the rims. Otherwise, the tire will come off the rim while driving at fast speeds.

Figure 7-2 *Printed NinjaFlex tires*

Print the rims with the hole for the bearing facing up as shown in Figure 7-4. This will better seat your bearing and the bearing covers will cover up any visible imperfections.

Try gluing the tire in small sections so it can be reused in emergencies. When you damage your tires, use a hobby knife or razor blade to cut the glued sections so as to separate the tire from the rim. Try not to cut the tire too much. One benefit of having a Trike is that tires come in sets of four, leaving you with a spare.

Figure 7-3 *Tires purchased from Amazon*

Figure 7-4 *Bearing, bearing covers, and screws*

Press the bearing in the tire and use three M3 x 8mm screws to secure the bearing plates over the bearing (Figure 7-5). Repeat this step with the second tire.

Figure 7-5 *Screwing on the bearing cover*

FRONT STEERING ASSEMBLY

Next, you'll put together the left and right front steering hub assemblies (Figure 7-6). You'll need two M3 x 20mm screws, and two M3 hex nuts, one for each side.

Figure 7-6 *Assembled steering hub assembly*

The printed parts required to assemble the front steering are the steering hubs (*Steering-Hub-Left.stl*, *Steering-Hub-Right.stl*) and the hub carriers (*Hub-Carrier-Left.stl*, *Hub-Carrier-Right.stl*).

Drill the screw through the parts and into the nut, spinning the drill until the nut sinks into the nut trap on the bottom of the hub carrier (Figure 7-6).

The steering hub should move freely; don't overtighten. If an M4 screw head is binding on the wheel due to a poor print, open up the hole up using a drill bit or knife.

Attach the completed left and right hub assemblies to the front wheels using one M4 x 12mm screw on either side. The M4 screw goes through the front of the wheel, through the inside of the 624zz bearing, and into the front of the steering hub.

Once it is in place and moving freely, you'll want to add the rim cover in order to finish the whole assembly as shown in Figure 7-7.

Figure 7-7 *Assembled hub and wheel with rim*

BACK WHEEL ASSEMBLY

To assemble the back wheel, you'll need one tire (printed or purchased) and one print of *Back-Rim.stl* and *Back-Wheel-Mount.stl*. You'll also need the NTM Prop Drive Motor, three M2.5 x 10mm screws, and four M3 x 10mm screws (Figure 7-8).

Before attaching the back wheel to the motor, clean the inside of the back wheel where the two meet so that the motor mounts flat to the wheel. Any malformed plastic inside will cause the wheel to mount and spin crooked.

Put the motor inside the wheel and screw them together using three M2.5 x 10mm screws (Figure 7-9). Ensure that the holes line up well so the screws will not cross-thread.

Figure 7-8 *Motor, back wheel, and screws*

Figure 7-9 *Motor attached to back wheel*

Take the wheel assembly and connect it to the back wheel mount (*Back-Wheel-Mount.stl*). Start by feeding the wires through the square hole next to the motor mounting plate (Figure 7-10). Feed the first wire through the hole and move it to one side so it will be out of the way of the next.

Figure 7-10 *Feeding wheel assembly wires through back mount*

The last connector is tougher to squeeze through the hole, so be careful not to ruin the wiring or connector on the motor. Once the wires are through, as seen in Figure 7-10, insert the motor shaft through the center hole on the motor mount plate (Figure 7-11).

 Be careful not to pinch or crimp the wires as you position the motor flush to the motor mount—it's a tight fit.

Make sure that the motor mount plate is debris free so the motor will mount straight. If you used support material it will take a little more work, but it's worth taking the time to do it right.

Figure 7-11 *Motor shaft inserted into motor mount plate*

Once the motor is in place, secure it using four M3 x 10mm screws (Figure 7-11). The holes should line up directly with the motor.

 Make sure the screws are not cross-threading as you put them in.

FRAME ASSEMBLY

When the large frame pieces (*Body-Back.stl* and *Body-Front.stl*) are finally done printing they are easily put together with just a few screws.

In addition to the frames, you'll need the servo, two M3 x 10mm screws, and four M3 x 16mm screws.

The tricky part is getting the frames to print out flat so they fit together well. The frame will want to curl up on the edges, and that makes it difficult to insert the shocks for the front wheels.

 Using the right tape or a heated bed will help in making the plastic stick to the build platform to prevent curling. Even with PLA, a little heat will help.

Figure 7-12 *Mounted servo*

Begin building the frame assembly by installing the servo into the servo mount (Figure 7-12). The servo should slide in between the mounts with a snug fit, and the servo brackets should fit tight against the mounts.

Insert two M3 x 10mm screws through the top two bracket holes of the servo mount to hold the servo in place (Figure 7-13). The wiring and servo horn should be on the front end of the servo facing the front of the car.

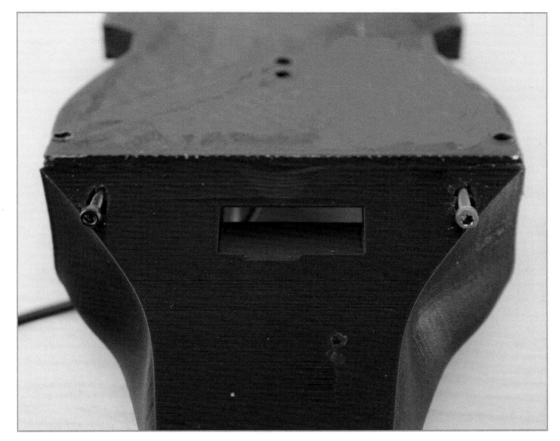

Figure 7-13 *Screwing into servo mount from under the frame*

After the servo is secure, attach the front and back bodies using four M3 x 16mm screws. Make sure your components are aligned so the screws do not strip out the plastic holes.

Attach the two screws on the top first, as they are easier to line up. You may need to use an extension on your drill in order to get the screw over the servo (Figure 7-14).

Figure 7-14 *Attaching the front and back bodies*

STEERING COMPONENTS

Now that the body is coming together, you'll begin assembling the steering components, then attaching them to the car.

Grab the two steering linkage prints (*Steering-Linkage.stl*), the top and bottom steering hinge plates (*Steering-Hinge-Bottom.stl*, *Steering-Hinge-Top.stl*), and the servo steering arm (*Servo-Steering-Arm.stl*). Attach the five parts together using three M3 x 10mm screws as seen in Figure 7-15.

Figure 7-15 *Attaching the steering components*

Start with adding the screws to the top steering hinge—it's the hinge with the smaller spacer on the bottom of the part.

One by one set the screws through the arm and linkage arms and into the bottom steering hinge. Be careful to not overtighten the part; it will limit the movement of the other parts. Everything should move freely.

 The steering linkage ends are different; the more rounded end goes into the steering hinge.

MOUNT THE STEERING SERVO

Bring the steering assembly to the frame and attach the servo steering arm to the servo as seen in Figure 7-16. You will need the hardware provided with the servo and a micro Phillips screwdriver.

Figure 7-16 *Attaching the servo steering arm*

I used the screw spacer because the screw was too long. Center the servo horn's travel, and find the three holes close together on the bottom that you can mount to. You can play with which one works best, but it seems to work best in the front or middle hole.

Figure 7-17 *Mount the steering hinge to the frame*

Now mount the steering hinge to the frame using one M3 x 16mm screw to attach the steering hinge mount to the frame from the bottom (Figure 7-17). Then one M3 x 25mm screw will go in from the bottom, up through the steering hinge and into the steering hinge, mount to hold the steering assembly in place.

INSERT THE SHOCK ABSORBERS

Gather the front wheels with the attached hub assemblies and print two copies of all of the shock absorber/stabilizer arm parts for the front end: *Large-Shock.stl*, *Small-Shock.stl*, *Stabilizer-Arm-Left.stl*, and *Stabilizer-Arm-Right.stl*, as seen in Figure 7-18.

You'll also need needle-nose pliers, superglue, and four M3 x 8mm screws.

Figure 7-18 *Shock absorber parts*

Insert the large shocks into the steering hubs and the small shocks into the stabilizer arms.

These parts can be difficult to insert depending on your printing tolerances, but it should be manageable with a couple of tools.

Figure 7-19 *Use needle-nose pliers to push shocks and stabilizer arms into place*

Use needle-nose pliers or an adjustable wrench to put shocks into place as shown in Figure 7-19. Then insert the shocks into the slots on the frame.

 If you're having trouble, you may need to remove burrs or the edge with a knife or some sandpaper. This will help the NinjaFlex to slide into the hole. The same burr-removal technique applies to typical plastic filament, if you've opted out of the NinjaFlex.

When you have all the parts together with the shocks in place, align the parts and add small dabs of superglue to keep the shocks from sliding parts out of alignment while in use.

Use very little glue, so that when the shocks or other parts break, they can be separated and replaced easily. Apply more glue to the large shock attaching the wheels to the frame than to the smaller shock.

Figure 7-20 *Begin inserting screws after gluing is complete*

As soon as the shocks and steering components are in place, you'll begin screwing them together with four M3 x 8mm screws.

Start by attaching the stabilizing arm to the hub carrier with the 8mm screws (Figure 7-20).

Then screw the steering linkage into the steering hub arm. Once again, the steering components need to be able to move freely, so don't tighten the screws too tightly.

Now that the steering components are all assembled, move them around to make sure that none of the parts are binding, then make any necessary adjustments.

ELECTRONICS

When hooking up the electronics on this trike, take great care with cable management and component placement. Otherwise you'll run into major problems when you attempt to drive the vehicle. Ensure that all plugs and wires are tucked away from the moving mechianical components before you try running the car.

INSERT THE SPEED CONTROLLER

First, insert the speed controller into the back body so the leads connecting to the motor are running through the back of the car. Slide the controller wiring though the hole on the underside of the body as shown in Figure 7-21. You'll need some heat shrink tubing and three M3 x 10mm screws.

Pull the wires attached to the NTM Prop Drive Motor through the hole in the back wheel mount so it looks like Figure 7-22.

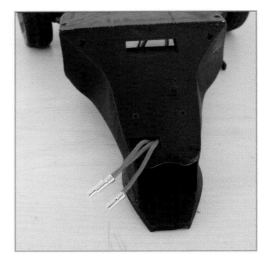

Figure 7-21 *Speed controller installed*

Figure 7-22 *Drive wires through back body*

CONNECT THE WIRES

You're going to connect the speed controller wires to the motor wires, but before plugging them in, cut three small 1″ sections of cable heat shrink tubing and slide a piece onto each of the speed controller wires.

Plug in the speed controller and motor connectors as you would any other electronic component: black to the negative, red to the positive, and yellow to the middle wire.

Slide the heat shrink over the connectors, covering the connections, but I highly recommend that you hold off on using the heat gun. Don't fully secure the wires until you have finished the build and can verify that the motor is spinning in the right direction. We'll cover that later in the build.

Figure 7-23 *Control board connected to the prop motor*

Push the partially shrink-wrapped connections and the speed controller wires up into the back body. You want them out of the way of the wheel, so the wheel mount sits flush with the back body. The wires should look like Figure 7-24.

Use three M3 x 10mm screws to mount the back wheel mount to the back body. Be very careful not to strip the plastic in the holes on the back body mount as the walls are thinner and it's very easy to strip the holes.

ADDING THE RECEIVER

First, run the servo wire under the servo steering arm and connect it to the first port on the receiver. The black wire faces the outside edge of the receiver as shown in Figure 7-25.

Figure 7-24 *Controller wires tucked away*

Figure 7-25 *Transmitter inserted*

Plug the speed controller into the receiver's second port, routing the wires in the same direction as the servo wires to keep them out of the way of the servo steering arm.

Wrap the loose cable and zip-tie the slack, but give yourself enough cable length to push the receiver and speed controller into the cavity of the back body.

You want the receiver's antenna to stick out of the back of the car, but ensure that you can still access the battery plug in the front of the cavity.

When the car is ready to be run, plug the battery in and insert it into the car as pictured in Figure 7-26.

Figure 7-26 *All components inserted*

PROGRAMMING AND USAGE

When getting ready to run the car for the first time, you must program the speed controller. To do this, plug in the charged battery and turn on the transmitter, holding the throttle all the way on.

The speed controller will beep twice. Let go of the throttle and it will beep three more times. After one long final beep the motor should be still and only move once the throttle on the transmitter is moved.

This is when you want to check the direction the motor is turning. If it is not turning in the right direction, switch the wiring on any two wires that connect the motor to the control board; this will fix this problem.

Once the direction is verified, use a heat gun to compress the heat shrink on the connectors to keep them from disconnecting.

After completing this step, all that is needed for future use is to turn on the transmitter then plug in the battery. You will not need to compress the throttle to program it any further.

Once the battery is connected and the servo and motor are properly moving in the correct directions and obeying the transmitter's commands, connect the front body frame using two M3 x 16mm screws (Figure 7-27). Then enjoy!

Figure 7-27 *Connect the front frame*

EXTRAS

There are many extra accessories that either increase the functionality or make for a fun car. I've included a few of them in the 3D printable file downloads.

If you want to go further, print and test out the wings (*Wing-Left.stl* and *Wing-Right.stl*) and the front bumper (*Bumper.stl*), both shown in Figure 7-28.

The bumper will protect the car from breaking when hitting something head on—which will eventually happen. The wings help the trike to make sharper turns at high speeds without flipping over.

Figure 7-28 *Bumper, wings, and screws*

To add the wings and the bumper, secure the parts with four M3 x 25mm screws on the bottom of the frame as depicted in Figure 7-29. The bumper screws into the front of the body, while the wings screw into the back.

Figure 7-29 *Attached wings, final bumper screws*

When attaching the wings, remove the 16mm screws currently holding the body together and replace them with the 25mm screws. On the end of the wings are holes for mounting something resilient, like a washer to weigh down the trike, making it even less susceptible to high-speed turnovers.

Figure 7-30 *Completed Inverted Trike RC, with wings and bumper attached*

Enjoy your new Inverted Trike RC (Figure 7-30)! Tweak it to make it your own and experiment to see how fast you can get it to go!

Skycam

by Brook Drumm

Building a robot from scratch has been a dream of mine since I was very young. There are so many robot projects and kits out there, I have no excuses. In fact, I have a few robot kits sitting in boxes in my garage right now. I just couldn't get excited about a line following robot or a simple tank. I even have the latest Lego EV3 gear, but have only played with it once. I just like doing things differently and attacking problems from a fresh perspective.

COST	
+ $140	

PRINT TIME	
+ 15 hours	

BED SIZE	
+ 6"x6"x6"	

ASSEMBLY	
+ 5 hours	

Skycam is a little robot that travels on a rope or string, can turn corners, and even has a little camera with a pan and tilt. [1] The whole thing can be remote-controlled from your phone or any browser—and it even streams live video.

This project was born out of the desire to make a robot that I can control from my phone. I had recently been playing with the Raspberry Pi and Google Coder, a neat bit of software that teaches the basics of web development. Using Coder, I was able to control an LED from my phone, so I knew controlling other hardware was possible. I enlisted friends to help me over the hardware and electronics hurdles, so I don't deserve full credit, but that's the beauty of the maker movement —a rich community of people ready to help!

My dream has come true. I built a robot. Not just *any* robot, but one that rides a monorail of string in the sky, runs its own web server, streams video, and is remote-controlled from my phone. *I love being a maker.*

---

1 Brian Roe helped by designing 90-degree and 45-degree 3D-printed parts to enable the elevated "monorail" of string to turn corners.

PARTS, TOOLS, AND FILES

FILES TO DOWNLOAD

To complete the *Skycam* project, you'll need to download the fabrication files and code (*http://bit.ly/1O6uoU8*) from the *Make: 3D Printing Projects* site (*https://github.com/Make3DPrintingProjects*).

PARTS		
Quantity	Description	PN/URL
1	Raspberry Pi B+ Wi-Fi Bundle	*http://bit.ly/1zTjCFJ*
1	8GB microSD card	Included with Raspberry Pi Wi-Fi Bundle
1	WiFi dongle	Included with Raspberry Pi Wi-Fi Bundle
1	Raspberry Pi camera	*http://bit.ly/1Fj1Q4X*
1	12" flex cable	*http://bit.ly/1Jjhkbq*
2	624 bearings	*http://amzn.to/1OjJs1I*
2	Slide switches	*http://amzn.com/B009752DE0*
2	Continuous rotation micro servo	*http://bit.ly/1Gom0Jt*, Part# FS90R
2	Micro servos	*http://bit.ly/1PdB3uo*, Part# HXT900
1	4-AA battery pack	*http://amzn.com/B000LFVFT4*
1	6-AA battery pack	*http://amzn.com/B000LFVFU8*
1	5V regulator	DigiKey: MC7805CT-BPMS-ND
1	100uF electrolytic capacitor	DigiKey: 493-1548-ND
1	2-pin male header	DigiKey: 3M9447-ND
4	1K resistors	DigiKey: CF14JT1K00CT-ND

PARTS

1	40-pin male header cut to size	DigiKey: S1011EC-40-ND
4	M3 hex nuts	*http://bit.ly/1Gp5Xhd*, Part# 90592A009
4	M4 12mm screws	*http://bit.ly/1ON9l4E*, Part# 91290A148
4	M4 washers	*http://bit.ly/1Gp61h9*, Part# 91166A230
2	Rubber bands	*http://amzn.to/1FiYJK8*
1	Female/female jumper wires (40 pack)	*http://bit.ly/1QqK6ca*

TOOLS & MATERIALS

3D printer	Printer filament
2.5 mm allen wrench	Needle-nose pliers
Wire cutters	Exacto knife
Soldering iron	Solder
Computer with microSD card slot	Micro USB cable
CAT5 cable	High test fishing line
Superglue	Zip ties (optional)

3D PRINTABLE FILES

Quantity	Filename	Quantity	Filename
1	*Skycam-bottom.stl*	1	*Skycam-camera-back.stl*
1	*Skycam-camera-front.stl*	1	*Skycam-camera-pan.stl*
1	*Skycam-camera-tilt.stl*	1	*Skycam-corners.stl*
1	*Skycam-left-wheel-mount.stl*	1	*Skycam-middle.stl*
1	*Skycam-pan-tilt-top.stl*	1	*Skycam-right-wheel-mount.stl*
1	*Skycam-top.stl*	2	*Skycam-wheel.stl*

ASSEMBLE THE TOP PLATE

Press the two 624 bearings into their pockets in the top plate (*Skycam-top.stl*), as shown in Figure 8-1.

ATTACH SWITCHES

Attach the two power switches to the top plate with four M3 x 8mm screws and nuts (Figure 8-2).

One switch is used to power the Raspberry Pi, the other is used to power the servos. I used two different switches so I could tell them apart, but I've modified the parts list and files to use two of the same type.

The pictures show two different switch types. The difference is purely aesthetic, not functional.

MOUNT THE SERVOS

Insert the two continuous rotation servos into the left and right wheel mounts (*Skycam-left-wheel-mount.stl* and *Skycam-right-wheel-mount.stl*).

Ensure the servo drive gear is lined up with the center of the 3D-printed vertical post, as shown in Figure 8-3. To operate properly, the center of the wheel needs to pivot directly above the 624 bearing when the servo post is mounted.

Screw each servo to the wheel mount using two M3 x 10mm screws.

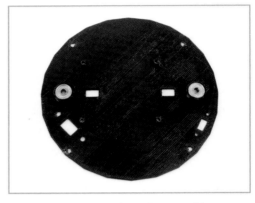

Figure 8-1 *Bearings inserted into top lid*

Figure 8-2 *Attached power switch*

If you'd like to learn how to convert regular servos into continuous rotation servos, I've written up (http://bit.ly/1HYQnX5) the process.

Insert one M4 x 12mm screw through each of the 624 bearings from the underside of the top plate.

The threads should be coming out of the top —the same side as the switches. The wheel assemblies will rotate on these screws.

As pictured in Figure 8-4, add two M4 washers to each of the screws threaded through the bearings.

These washers allow the wheel mounts to rotate freely on the center of the bearings.

Thread the ends of the screws into the left and right wheel mounts as shown in Figure 8-5. Tighten the screws until the connecton to the wheel mount is rigid.

After tightening, rotate the wheel mounts. They should be able to rotate freely on the screw post inside the bearing.

Figure 8-3 *Servo in right wheel mount*

Figure 8-4 *Screw and switch orientation*

Figure 8-5 *Pre-tightened wheel mounts*

ADD THE WHEELS

Press the wheels (*Skycam-wheel.stl*) onto the servo drive posts. Secure the wheels with the small self-tapping screw that came with the servos.

Wrap a small rubber band in the valley of the wheel hub. The rubber band will help the wheel grip against your paracord as depicted in Figure 8-6.

Route your servo wires through the square holes adjacent to the mounted wheels (Figure 8-7). Leave enough slack on the wires to allow the wheels to rotate 45 degrees in either direction.

Figure 8-6 *Wheels screwed to servos*

Figure 8-7 *Routed servo wires*

RASPBERRY PI SD SOFTWARE SETUP

It turns out that the Raspberry Pi is a great platform for building robots! You can even create your own custom user interface and stream video live to your browser.

To get these features up and running, you'll need the *Google Coder* software platform and the *Pi-Blaster* software library that communicates with the servos. You'll also need *MJPG-Streamer* to stream video from the Pi's camera.

INSTALL GOOGLE CODER FOR RASPBERRY PI

For the setup and installation of the required programs, you'll need to connect your Pi to your router directly using a Cat5e Ethernet cable. But plug the Wi-Fi adapter into your Raspberry Pi anyway.

Do not connect your Pi to your computer/router with the Ethernet cable just yet.

Point your computer's browser to the Google Coder (*http://googlecreativelab.github.io/coder*) project and select the "Wired" instruction set button, as shown in Figure 8-8.

To install Coder and get connected, complete steps 1 and 2 as shown in the screenshot:

1. Download the coder installer, unzip, and run the installer as directed.

2. Insert your SD card, attach your Ethernet cable to a network connection, and power your Pi.

3. After your Pi powers on, head to *http://coder.local* using the Chrome browser and connect to your Pi.

All set!

During step 3, your browser will warn you that "Your connection is not private." Click "Advanced," and then "Proceed to coder.local."

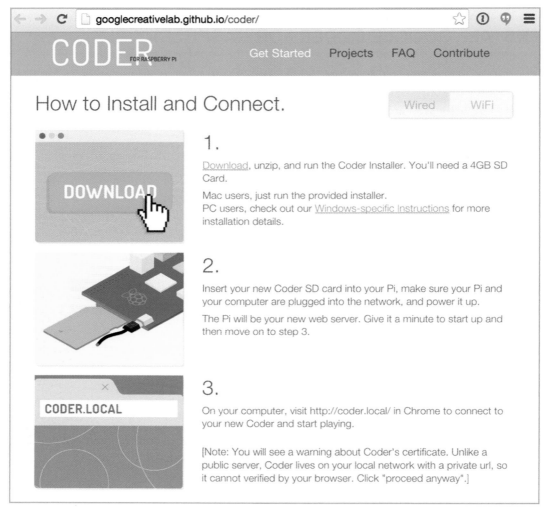

Figure 8-8 *Google Coder "Wired" selected*

GOOGLE CODER

Coder will prompt you to provide the password you set when initially setting up Coder. Once you're logged in to your Pi and are on the Coder home screen, you'll set up the wireless capability.

In the top-right corner of your browser window you'll see a gear icon. Click it and select "Wifi Setup," then follow the onscreen instructions.

Disconnect the Ethernet cable from the Pi so that you are now running on the WiFi network. Reboot your Pi.

CODER INTERFACE

Log back in to your Pi by navigating to *http://coder.local* in Chrome. Ensure that your wireless connection is working correctly. Nice job! You're halfway there.

Explore the Coder environment (Figure 8-9). You'll be able to inspect and even edit the code right there in your browser window (Figure 8-10)!

Figure 8-9 *Coder enviroment*

Figure 8-10 *Editing code in Coder*

INSTALL PI-BLASTER ON YOUR RASPBERRY PI

 You'll need to hit the Enter key after typing each command.

1. To install Pi-Blaster, the first thing you'll need is a network scanning program so you can identify the IP address of your Pi. I used LanScan (*http://apple.co/1MYYcSG*) from the Apple Store (Figure 8-11), but any network scanning software you'd like to use will work. Windows has lots of options as well.

2. Once you have your network scanner, search your network for the name "Edimax Technology Co. Ltd." This is your Pi, and you'll want to take note of the IP address associated with this name. The address of my Pi was `192.168.1.115`, but yours could be different.

3. Open the terminal to access the command line on your computer and begin by typing the following line, inserting your own IP address:

   ssh `192.168.1.115` **-l pi**

4. You'll be asked if you want to continue (you do). Type:

   yes

Figure 8-11 *LanScan*

5. When you're prompted for a password, enter your Coder password.

6. Next, to download the Pi-Blaster software, enter the command:

   wget https://github.com/sarfata/pi-blaster/archive/master.zip

   In the terminal window, you'll see confirmation that the *master.zip* file is downloading and a progress bar. This may take a moment.

7. Once the download is complete, as shown in Figure 8-12, you can unzip the archive by entering:

   unzip master.zip

8. To move into the Pi-Blaster folder you created when you extracted the *ZIP* file, type:

   cd pi-blaster-master

   Your command prompt should change to **pi@coder: ~/pi-blaster-master$**.

Figure 8-12 *Downloading the files*

9. You'll now need to install the files in the *pi-blaster-master* folder. To do this enter:

   sudo apt-get install autoconf

   You'll be asked if you want to continue (type "Y" for Yes):

   Y

10. To build the software, you'll begin by typing the command:

    `./autogen.sh`

    This will take a moment to run.

11. Then type:

    `./configure`

12. Finally, compile the code:

    `make`

13. Then install Pi-Blaster:

    `sudo make install`

Pi-Blaster is now ready to run on your Pi!

 Do not close your terminal window or exit from the ssh session with your Pi yet, as there's more to be done. You still have to enable streaming so you'll be able to view what your Pi's camera is "seeing"!

INSTALL MJPG-STREAMER AND ENABLE YOUR CAMERA

1. To enable the Camera function on your Pi, type:

   `sudo raspi-config`

   The Raspberry Pi Software Configuration Tool GUI (graphical user interface, as opposed to the command prompt) will appear.

2. Using the arrows on your keyboard, move down the option list until option #5, "Enable Camera," is highlighted, then press Enter.

3. The system will ask you if you are sure (you are). Use the cursor keys to highlight "Enable," and then press Enter.

4. You'll return to the main menu screen. Press the right arrow key once, and the "<Finish>" option should now be highlighted. Press Enter.

   You'll be prompted that you need to reboot to enable the changes you just made. Reboot your Pi now:

   `sudo reboot`

5. Once your Pi has rebooted, log back in to the Pi over the ssh connection as you did earlier (see "Install Pi-Blaster on Your Raspberry Pi" step 3).

6. You'll need to update the Pi so that it knows the location of all of the files and data you're about to load onto it:

`sudo apt-get update`

Files being downloaded will fly by on the terminal window. This will take a few minutes.

 After each command entered during steps 7 to 11, you'll be prompted for a yes or no answer. Answer "yes" to all queries, or the installation will be aborted and you'll have to start over.

7. When you get the command prompt back, the Pi has been updated, but you still need some additional software, so type:

`sudo apt-get install subversion libjpeg8-dev imagemagick libav-tools cmake`

8. Now you'll need to download the MJPG-Streamer software by cloning the GitHub repository where the files are stored:

`git clone https://github.com/jacksonliam/mjpg-streamer.git`

You'll end up with a folder called *mjpg-streamer/mjpg-streamer-experimental*.

9. To move into the *mjpg-streamer* folder, so you can build the software, type:

`cd mjpg-streamer/mjpg-streamer-experimental`

10. Now compile the software:

`make`

11. Then enter:

`sudo make install`

12. You need to tell the MJPG-Streamer that you're using a Raspberry Pi camera:

`./mjpg_streamer -i "./input_raspicam.so -fps 5" -o "./output_http.so -w ./www -p 8090"`

13. Test that the streamer and camera are working together by opening an Internet browser window on a separate computer, then navigate to the address **http://_CODER-IP_:8090**, replacing CODER-IP with your Pi's IP address; for instance, I would enter *http://192.168.1.115:80*.

14. You should now see a page that looks like Figure 8-13. In the second picture, under the flower, if you see a snapshot of what your camera sees, then the setup is working.

15. Move back to the terminal session connected to the Pi, hold down the Control key, and press the X key to stop MJPG-Streamer from running. To configure the streamer to start automatically whenever your Pi is powered on, it needs to be closed. First you'll need to move to your settings folder:

```
sudo nano /etc/rc.local
```

The *nano text editor* will open the file *rc.local*, displaying something similar to the screenshot in Figure 8-14.

16. You'll need to add a few lines of code to the bottom of the *rc.local* file. Place them after the `fi` and just before the `exit 0` at the very bottom, above the red bar (shown for placement; it won't appear in the text editor) as shown in Figure 8-14.

Figure 8-13 *MJPG-Streamer connected to camera*

 You'll need to use your arrow keys to navigate down to below `fi` and above `exit 0` to type or paste in the code. Your mouse won't work.

Here's the code to add:

```
cd /home/pi/mjpg-streamer/mjpg-streamer-experimental/./mjpg_streamer -i
"./input_raspicam.so -fps 5" -o "./output_http.so -w ./www -p 8090"
```

17. Once you have added the two lines to the config folder, hold down the Control key and press X to exit. You'll be asked if you want to save the changes; type Y for yes and press Enter. You'll then be given the option to change the folder name; do not change this as we want to overwrite the config folder to have the new settings, so just press Enter to continue.

Figure 8-14 *The red bar appears after the added code*

18. *Whew!* That was a lot. (Huge thanks to friend and code wizard Mick Balaban and Nick Ernst, Printrbot Engineer, for getting me sorted, writing these tutorials, and collaborating on fun projects like these!)

Now reboot your Pi and you are ready to go!

```
sudo reboot
```

ASSEMBLE THE MIDSECTION

MOUNT THE RASPBERRY PI

Once you have verified that everything is working over WiFi, use four M3 x 8mm screws to mount the Raspberry Pi to the underside of the top lid.

Connect the top lid assembly to the middle case (Figure 8-15) with four M3 x 8mm screws.

Figure 8-15 *Pi attached to top lid*

 Figure 8-15 shows the Raspberry Pi mounted in place, along with the two "business" ends of our switches. As mentioned previously, I used two different switches to tell them apart. You can use two of the same switches if you prefer.

ADD THE ENDSTOPS

The middle case has two flanges for endstops, also called micro switches. The endstops are used to trigger actions if they hit a wall or obstacle.

Figure 8-16 *Endstops*

Secure the endstops with four M2.5 x 10mm screws with the switches facing out. Route the wires through the holes in the side under the flanges as shown in Figure 8-16.

You'll wire everything up once all the mechanical pieces are in place.

CAMERA ASSEMBLY

The Skycam camera assembly hangs off the bottom lid of the robot. It uses a Raspberry Pi camera and two servos to pan and tilt while it streams the video to your browser.

ADD THE PAN/TILT SERVOS

Press one servo straight into the middle of the camera pan disc (*Skycam-camera-pan.stl*), ensuring that the servo shaft is positioned in the very center of the disc.

Use two M3 x 8mm screws to secure the servo.

Press the bottom of the camera tilt servo into the recessed area provided on the camera pan disc, directly next to the face-down servo. The servo shaft and wires should be on top (Figure 8-17).

Figure 8-17 *Servo inserted into pan disc*

The tilt servo is a press-fit affair, so consider taping the two servos together if needed for stability.

Press one of the servo horns into the recess on the tilt arm (*Skycam-camera-tilt.stl*) and secure it with a dab of superglue.

Press the tilt servo cap (*Skycam-pan-tilt-top.stl*) onto the tilt servo (Figure 8-18).

The tilt servo cap is a nonfunctional part. I just like the way it visually ties together the two servos into one block.

Figure 8-18 *Tilt servo cap added*

Optionally, the wires can be routed and pinned with a zip tie. Just make sure the servo wires have enough slack to allow the camera to pan and tilt.

ASSEMBLE THE CAMERA CASE

Use a dap of superglue to join the tilt arm (with the embedded servo horn) to the side of the back camera box (*Skycam-camera-back.stl*), matching up the shapes (Figure 8-19).

Figure 8-19 *Gluing the tilt arm to the back camera box*

 The camera box is an existing design, used here with permission. I wanted to keep it untouched like the original part, so superglue will do the trick.

Mount the Raspberry Pi camera on the tiny pins in the back camera box (*Skycam-camera-back.stl*) as shown in Figure 8-20.

Then add the camera cover (*Skycam-camera-front.stl*). It snaps on with a light press (Figure 8-21).

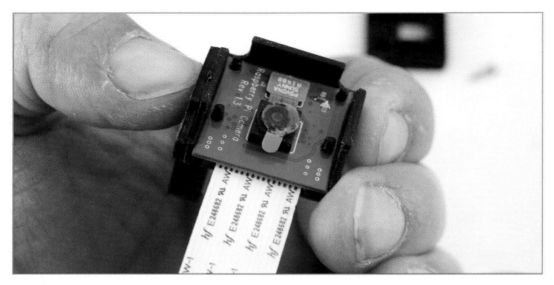

Figure 8-20 *Pi camera mounted in box*

Figure 8-21 *Camera cover in place*

MOUNT THE CAMERA TILT ARM

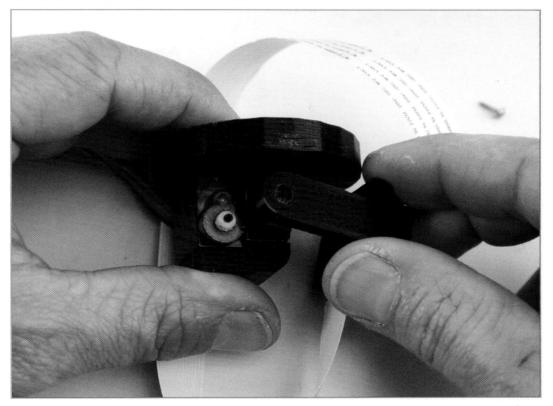

Figure 8-22 *Mounting the camera arm*

Mount the camera tilt arm on the servo, but don't secure it with the screw just yet (Figure 8-22). If the tilt arm servo and arm are not positioned correctly, the tilt arm will hit the camera pan disc.

Hold off from adding the screw until later in the project. Once you turn on the servos, you'll be able to verify that the tilt arm servo is moving properly.

ATTACH THE CAMERA TO THE BOTTOM PLATE

Push the servo horn shown in Figure 8-23 through the back of the bottom lid (*Skycam-bottom.stl*) and secure it with a few drops of superglue.

Thread the camera cable and servo wires through the slot in the back lid, leaving plenty of slack.

Figure 8-23 *Threading camera and servo cables*

Mount the entire camera assembly to the bottom lid using the tiny servo screw (Figure 8-24).

It's likely you'll end up repositioning the camera mount angle, but this servo doesn't have a hard stop like the tilt arm in Figure 8-22 does. You won't do any damage if it's pointing in the wrong direction.

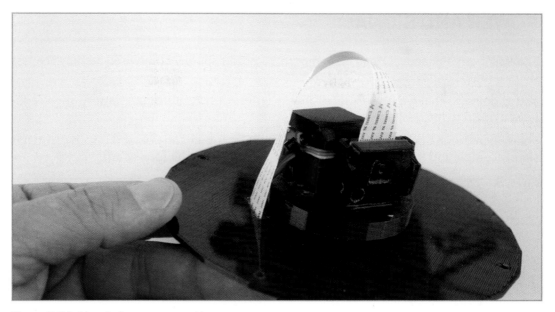

Figure 8-24 *Mounted camera assembly*

TEST FIT

Test-fit the bottom lid with the middle case and bottom plate assemblies to avoid any setbacks later. The bottom lid attaches with four M3 x 8mm screws.

After the test fit, you'll hook up the electronics and test the software to make sure everything moves properly before closing it up.

CONNECT PI CAMERA

You should connect the camera ribbon cable to the Pi now, so you'll be ready to test the video streaming when the project software boots up.

ELECTRONICS!

The brains of the Skycam can be repurposed for other projects. You can use what you learn here to design your own 3D-printed remote-controlled car or tank with a streaming camera!

To keep this project ultra-flexible, you'll be making your own power and signal boards. Hang in there if this is your first time hacking together one-off electronics. When it's done, you'll be able to step back and see that you've tapped into a new superpower!

POWER AND SIGNAL

The Raspberry Pi needs a clean and smooth 5V power source, otherwise you may see less than desirable behavior.

You're going to make a power board that takes in 8.4–12VDC from a battery, then regulates it down to the desired 5V for your Pi.

The switch doesn't power the Pi itself; it switches the ground line on and off. Our switch is simply completing the circuit (power on), or opening the circuit (power off).

BATTERY CHOICES

The power source you choose for this project can be anything that delivers 8.4 to 12VDC to the power board.

If you want to use your own battery, you could use a 2- or 3-cell LiPo battery commonly used with multicopters, or two 18650 Li-Ion rechargeable batteries. You could just use the recommended 6-AA battery pack from the parts list.

Whichever direction you choose, ensure that your battery pack and your power board have compatible connectors so that you can link them together!

I'm showing the electronics outside the case so you can see everything clearly. Both boards are made using small pieces of perforated board and use three pin headers for wire connections. Schematics and pictures are provided.

POWER/REGULATOR BOARD

To assemble the power/regulator board, you'll need the 100uF electrolytic capacitor, 5V regulator, a 3-pin header, solder, and a soldering iron.

In Figure 8-25, the white wire is the battery pack's voltage in, black is ground, and red is the 5V out of the regulator to the Raspberry Pi.

Figure 8-25 *Power/Regulator board*

Look over the schematic pictured in Figure 8-26 to clarify the power/regulator board circuit.

You'll need to solder the board's components together underneath the protoboard, as shown in Figure 8-27.

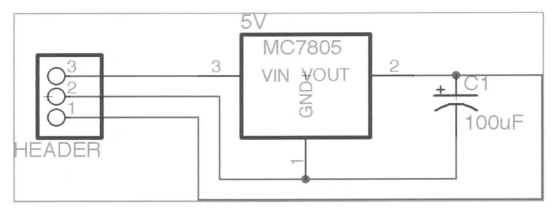

Figure 8-26 *Power/Regulator board schematic*

As you can see in Figure 8-27, I employed a blue "EC3"-style connector to attach the battery pack to the power/regulator board and switch. A standard hobby servo cable was used to connect power and ground to the Raspberry Pi.

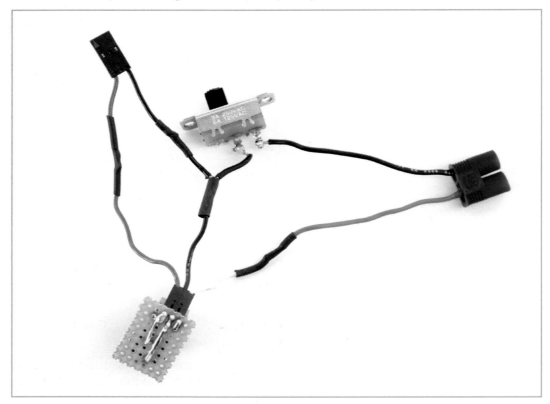

Figure 8-27 *Wired power board*

Hook up your 6-AA battery pack; then, using a multimeter, measure the output of your power/regulator board to ensure that you're getting 5V out.

If you're not getting 5V, check your connections and retest before connecting this board to your Raspberry Pi.

Once you've made and checked the necessary connections, don't forget to cover the board in electrical tape so you don't short anything out (Figure 8-28). There's not much room inside the Skycam case and it's likely that components will rub up against each other.

Now onto the signal board!

Figure 8-28 *Covering the power board in electrical tape*

SIGNAL BOARD

The signal board uses 3-pin headers to power and control the servos and connect them to the Raspberry Pi. It also connects the Skycam endstops to the Pi, so Skycam knows when it has hit a wall.

The signal board has its own, separate 4-AA battery pack, and will use the second switch you installed earlier (Figure 8-27) to turn power on and off.

You'll create a harness similar to the one you made for the power board to connect power to the signal board by soldering the 3-pin headers and resistors through the bottom of the perfboard as shown in the schematic (Figure 8-29).

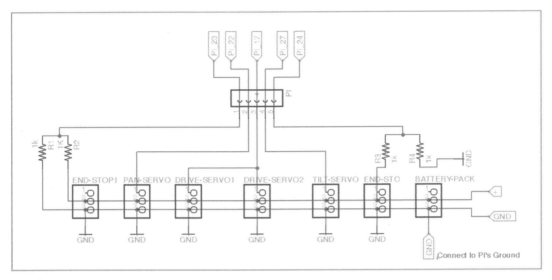

Figure 8-29 *Signal board schematic*

Starting from the battery pack, the positive (red) wire is connected directly to the signal board's positive input.

The negative (black) wire from the battery pack goes to one of the two ground terminals that are not connected to the servos or endstops. You can see these two exposed ground connections on the right side of Figure 8-29. After connecting to one of the grounds, hook up the other end to the switch.

Figure 8-30 *Perfboard signal "patch" board*

Next, you'll connect another black wire from the middle terminal of your switch to the remaining GND input on the signal board. The switch opens or closes the circuit, turning the power on and off.

CONNECT THE PI

Once you have finished soldering your boards, connect your jumper wires from the signal board to your Pi, as shown in Table 8-1.

Table 8-1 *Raspberry Pi connections*

Connection	Raspberry Pi pin
Endstop 1	RPi pin 16 → GPIO23
Pan servo	RPi pin 15 → GPIO22
Drive servos	RPi pin 11 → GPIO17
Tilt servo	RPi pin 13 → GPIO27
End stop 2	RPi pin 18 → GPIO24
Ground	RPi pin 14

Wrap the signal board in electrical tape to protect it from shorting; as mentioned previously, there's not much room inside the Skycam body.

Figure 8-31 *Signal board connected to Pi*

Now that you have a power board and a signal board, plug everything in, and turn on the Pi and the power to the servos. It's time to test out the software controls!

TEST THE SOFTWARE CONTROLS

Using a browser on a computer, connect to the Pi as you've done previously.

Once you are connected to the Pi and have set up Coder for Raspberry Pi, download my Skycam Coder (*https://github.com/Make3DPrintingProjects/Skycam*) project.

After downloading, upload the Skycam Coder project to Google Coder and it will appear in the menu.

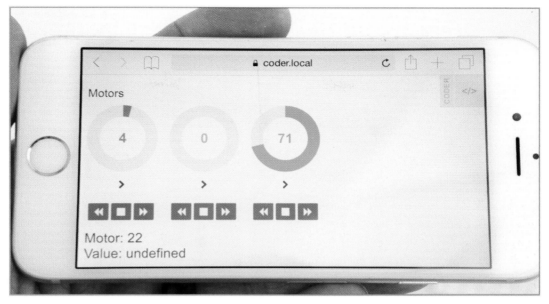

Figure 8-32 *Skycam Coder project for Raspberry Pi in a mobile browser*

Remove your tilt arm to allow the servo to center (this is why you didn't screw it in earlier). Use the Coder interface to drive the Skycam and pan and tilt the camera servos (Figure 8-32).

Test the tilt to get the right rotation before attaching the tilt arm and securing it with a screw.

 The streaming video should also be live at this point.

Assuming everything works perfectly, tuck in the wires and attach the bottom lid back with four M3 x 8mm screws.

Build your skyway with high test fishing line. You'll need to tie off to very secure places—I used interior door hinges in my house.

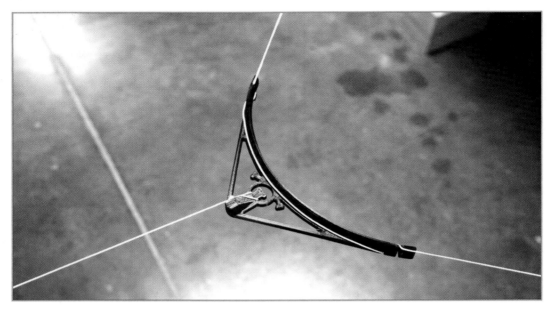

Figure 8-33 *Skyway corner*

You can get creative with the skyway and even build in turns using 3D-printed corners as seen in Figure 8-33. Using a corner requires an extra length of string attached to the back to "pull" the corner into position.

I've included a 90-degree turn and a 45-degree turn with the other 3D printable files.

Make sure the fishing line exits the corners right at the top so your Skycam wheels can make the transition. You'll need to keep the skyway string fairly level since there is not a lot of torque in the little micro servos.

Best of luck and have fun!

Figure 8-34 *Skycam rolling along*

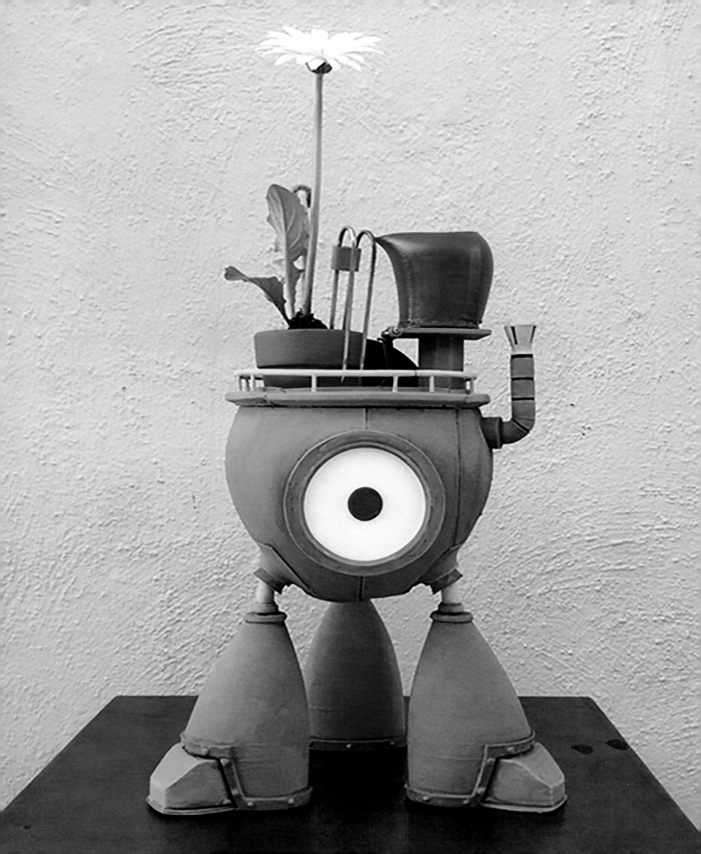

Chauncey: The Wrylon Robotical Flower Care Robot

Written by John Edgar Park
Design and Illustration by Barry McWilliams

Robots and flowers. Both are beautiful. Robots are endearing and highly responsible, while flowers bring peaceful joy, yet tend to die when ignored. Enter Chauncey, the Wrylon Robotical Flower Care Robot. This lovely little fellow will proudly display your favorite potted flower and ensure that she gets proper care.

COST
+ $90
PRINT TIME
+ 40 hours
BED SIZE
+ 6"x6"x6"
ASSEMBLY
+ A weekend

Due to certain regrettable events, full-scale production of the FLORABOT 3L-1G model, a.k.a. Chauncey, ceased in 1913. However, do not despair. With the never-before-published plans shown here, for the first time, you can build your own fully functional, 3D-printed Chauncey.

All assembly required. Certain restrictions apply. See https://github.com/Make3DPrintingProjects for details. Maker Media is not responsible for any robot uprisings, botanical or otherwise.

FILES, PARTS, AND TOOLS

FILES TO DOWNLOAD

To complete the *Chauncey: The Wrylon Robotical Flower Care Robot* project, you'll need to download the fabrication files and code (*http://bit.ly/1Gfkec6*) from the *Make: 3D Printing Projects* site (*https://github.com/Make3DPrintingPro jects*).

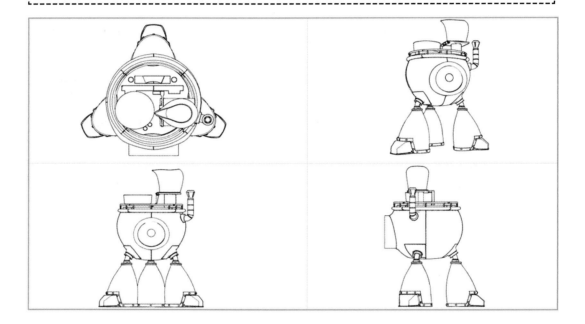

3D MODELING

The Chauncey model was initially modeled in Modo, by Barry, to make the translation from his drawings to 3D. I then brought those models into Maya to refine and cut for 3D printing. I modeled other parts (such as the deck) in Rhino.

PARTS		
Quantity	**Description**	**Part number**
1	Arduino UNO	Adafruit 50
1	Adafruit Motor Shield v.2	Adafruit 1438
1	WingShield Industries ScrewShield *OR* MakerShield Prototyping Shield (you only need one)	ScrewShield: Adafruit 196, MakerShield: MakerShed MSMS01
2	1/8" outside diameter brass tubing, 12" in length *OR* you can use two 3" stainless steel nails as a substitute	Brass tubing K&S Metals 8127 (*http://bit.ly/1FGvVNQ*)
1	Gearhead DC motor with 33:1 ratio, 24 VDC gear motor with turntable	allelectronics.com (*http://bit.ly/1L2JI1V*) DCM-351
1	Duratrax Power Kit *OR* any comparable 7.2V 1500mAh NiMH battery and charger	Duratrax DTXP4615
1	4 X AA battery pack with 2.5mm barrel jack for Arduino	Sparkfun 09835
4	AA batteries	
1	10mm green LED	
1	10mm red LED	
2	10kΩ resistors	
1	220Ω resistor	

TOOLS & MATERIALS	
3D printer	Printer filament in various colors (choose colors in advance, or paint the 'Bot later)
Soldering iron, solder, and exhaust fan	Drill with 3/32" bit
Wire cutter/stripper	Size 0 Phillips screwdriver
Dremel or other rotary tool for filament welding	Multimeter
Safety goggles	One 3" terra cotta flower pot, lined with a plastic bag, filled with soil and a lovely flower
Three 1/4"-20 x 3/4" screws and nuts for mounting motor and watering base	Quart-size zip-top plastic bag
22AWG stranded wire	Cyanoacrylate glue (superglue)

OPTIONAL TOOLS & MATERIALS	
Pliers for bending heated brass tubing	1-1/2" outside diameter bending form, such as a hammer head
Vise for securing the brass tubing	Propane torch
Molex or other interconnects	White spray primer
Acrylic craft paint	Small paintbrushes
1/8" Baltic Birch plywood for decking	Wood stain
Laser cutter (preferred), jigsaw, or scrollsaw for cutting the wood deck	

3D PRINTABLE FILES		
Quantity	**Filename**	**Body part**
1	*fBot_allParts.stl*	All body parts in a single file
3	*fBot_foot.stl*	Feet and legs
3	*fBot_bend.stl*	Feet and legs
3	*fBot_legSocketCenter.stl*	Feet and legs
3	*fBot_legSocketLeft.stl*	Feet and legs
3	*fBot_legSocketRight.stl*	Feet and legs
3	*fBot_lowSocket.stl*	Feet and legs
1	*fBot_bodyBL.stl*	Body
1	*fBot_bodyBR.stl*	Body
1	*fBot_bodyFL.stl*	Body
1	*fBot_bodyFR.stl*	Body
1	*fBot_eyeTube.stl*	Eye
1	*fBot_iris.stl*	Eye
1	*fBot_pupil.stl*	Eye
1	*fBot_deck.stl*	Deck
4	*fBot_railing.stl*	Deck
1	*fBot_pipeBottom.stl*	Stovepipe
1	*fBot_pipeTop.stl*	Stovepipe
1	*fBot_wateringCan.stl*	Watering contraption
1	*fBot_waterBase.stl*	Watering contraption
1	*fBot_waterLid.stl*	Watering contraption

3D PRINTABLE FILES		
1	*fBot_waterLever.stl*	Watering contraption
1	*fBot_wheelNub.stl*	Motor
1	*fBot_probeSpacer.stl*	Probes
1	*fBot_xx_woodDeck.svg*	Optional upgrade
1	*fBot_xx_motorBracket.stl*	Optional upgrade

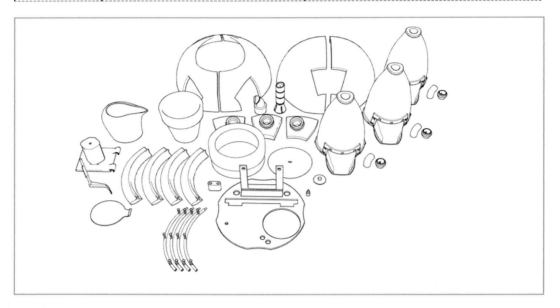

FABRICATION PHASES

Building your own Wrylon Robotical Flower Care Robot can be a fun, fulfilling, and useful project. Just as it was when the original assembly lines were humming at the Wrylon factory, creating the robot is divided into two major phases:

3D Printing and Fabrication
> Where you will print many of the robot's body parts, assemble them, and optionally paint the robot.

Electro-Mechanical Assembly
> Where you will create the sensors, brains, and actuator assembly, and imbue the robot with intelligence and personality.

PRINT AND ASSEMBLE THE LEGS AND FEET

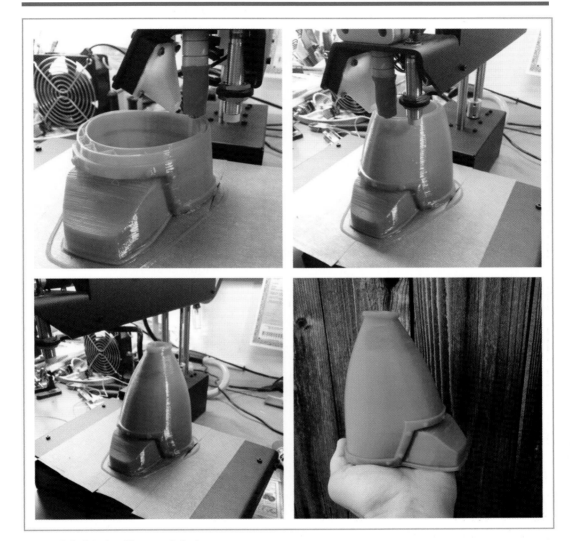

Figure 9-1 *Printing Chauncey's feet*

BIG CREATIONS FROM SMALL PRINTERS

Chauncey's parts have been carefully designed to print on a small desktop 3D printer with a build area of roughly 6"x6"x6". The assembled final robot is much larger than one that could be printed in a single pass on most printers.

Print three each of the foot (*fBot_foot.stl*), lower socket (*fBot_lowSocket.stl*), and upper leg (*fBot_legBend.stl*).

Assemble the legs by pushing the lower sockets into the tops of the feet, and the bends into the lower sockets, as seen in Figure 9-2.

Figure 9-2 *Attaching low sockets and leg bends*

PRINT THE BODY

In order to create a body larger than the build platform of most printers, Chauncey's body is printed in four sections and later assembled. Print one copy of the *fBot_bodyFL.stl*, *fBot_bodyFR.stl*, *fBot_bodyBL.stl*, and *fBot_bodyBR.stl* files (Figure 9-3).

Before joining the body parts, print the three leg sockets (*fBot_legSocketCenter.stl*, *fBot_legSocketLeft.stl*, and *fBot_legSocketRight.stl*), and the eye tube (*fBot_eyeTube.stl*).

 In the photos you will note that the front left body section is blue and white, due to a mid-print filament switch.

Figure 9-3 *Printed body sections*

ASSEMBLE THE BODY

In order to assemble the body parts we'll use a dual-bonding technique of gluing and friction welding.

The glue creates a bond to hold the parts together temporarily, but it's the friction welding that creates the real strength, since the friction-heated PLA plastic bonds the parts together. This method also has the benefit of filling in any gaps between parts.

GLUE AND FRICTION-WELD THE BODY

To begin assembling the body, place a small amount of superglue on the clean surface of one of the two parts to be joined as in Figure 9-4.

Press the parts together for 30 seconds as shown in Figure 9-5.

Figure 9-4 *Adding superglue*

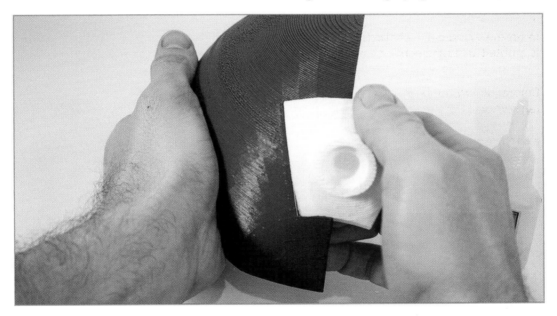

Figure 9-5 *Press the parts together*

FRICTION-WELDING TECHNIQUE

Chuck the free end of a PLA filament spool into your rotary tool and tighten the chuck. Then use diagonal end cutters to snip off the PLA filament—leaving about 1/2" of filament protruding from the tip.

Turn the rotary tool up to a 25,000–30,000 RPM setting. Moving in small circles, push the filament tip into the seam or gap you wish to weld, moving back and forth and overlapping across to both sides of the seam (Figure 9-6). Press hard enough that you see the PLA melt a bit as it heats up. Figure 9-7 shows the small circular motions embedded in the plastic seam.

Figure 9-6 *Using the rotary tool*

 Use eye protection to avoid injury from flying plastic bits!

The filament bit will get used up as you work. Turn off the tool, waiting for it to stop spinning. Loosen the rotary tool chuck, pull out another 1/2" length, retighten, and turn the tool back on to continue welding. Repeat this until you have to refill the tool with a new length of PLA from the spool.

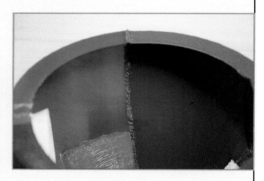

Figure 9-7 *Friction-weld closeup*

Welds can take 10 to 20 seconds to fully cool. You can take advantage of this by adjusting the fit of some parts while the weld is still warm.

ASSEMBLE THE LEG SOCKETS

Once the body welds are made, glue the center leg socket to the back right panel. The glue will help with holding the pieces in place when you later friction-weld the seams.

Let this part dry, then fit and glue the back right and back left body quarters to the socket. Once these have dried you'll use the friction-welding technique to permanently bond the parts as shown in Figure 9-8.

Proceed in this manner, gluing and welding the left leg socket to the front left body quarter, and the right leg socket to the front right body quarter.

Figure 9-8 *Center leg socket welds*

These two quarters will be closed around the eye tube before gluing and welding (Figure 9-9).

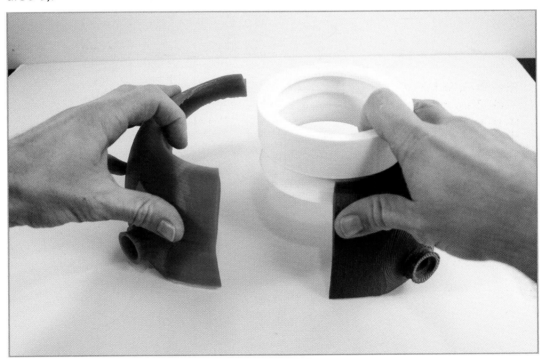

Figure 9-9 *Fitting the front body quarters to the eye tube*

Some small tack welds (Figure 9-10) will be enough to keep the eye tube in place.

Figure 9-10 *Small tack welds*

ADD THE EYE

Print the *fBot_iris.stl* and *fBot_pupil.stl* parts and then fit the iris into the eye tube from the back, and the pupil into the iris from the front (Figure 9-11).

Figure 9-11 *Attaching the eye*

COMPLETE THE DECK RIM, RAILING, DECK, AND STOVEPIPE

The top of the 'Bot has a deck rim running all the way around it to hold the deck in place. Print four copies of *fBot_rim.stl*, then glue and weld them to the top of the 'Bot's body.

If there are any gaps between sections, these can be filled in with friction welds (Figure 9-13, for that time-worn, hard-working robot look.

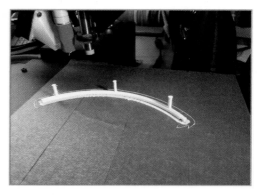

Figure 9-12 *Printing a section of safety railing*

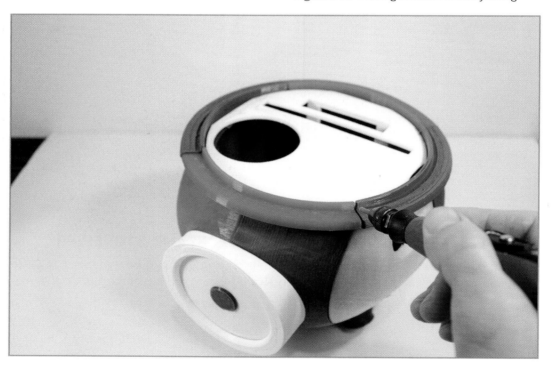

Figure 9-13 *Friction-welding the deck rim*

Print four copies of the safety railing model, *fBot_railing.stl* (Figure 9-12), and then join them together, as in Figure 9-14.

The railing will go on top of the deck rim (Figure 9-15); you can glue it on before or after painting.

Figure 9-14 *Joining the four railing sections*

Figure 9-15 *Railing added*

Print the deck. It rests on the body opening, and lifts off for adding electronics and wiring, and for changing or recharging batteries (see Figure 9-16). The motor, flower pot, indicator lights, and moisture probes all fit in or mount on the deck.

Print the two parts of the stovepipe, and then glue them together.

Once dry, push the stovepipe assembly into the port hole in the body to check the fit (Figure 9-17). We'll be priming and painting it separately, then reinserting it later.

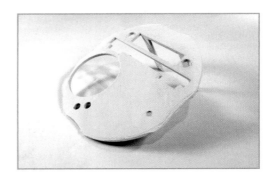

Figure 9-16 *The deck*

Figure 9-17 *Assembled stovepipe*

TEST-FIT THE LEGS

Figure 9-18 *Fitting the legs*

The legs of the 'Bot can be placed into the sockets without permanently adhering them (Figure 9-18), just to check the fit (Figure 9-19). We'll assemble it again later after priming and painting.

Figure 9-19 *Chauncey standing on all three feet*

CREATE THE WATERING CONTRAPTION

Next, we'll create the watering contraption. Print the watering can (*fBot_wateringCan.stl*), the base (*fBot_waterBase.stl*), the lid (*fBot_waterLid.stl*), and the lever (*fBot_water-Lever.stl*).

Glue and friction-weld the watering can to the lid (Figure 9-20), and the lid to the lever. The hinge pin snaps into the base so the watering can tips when the lever is pushed (Figure 9-21).

Figure 9-20 *Watering can bonded to lid/lever*

Figure 9-21 *Watering can tips when lever is pushed*

MOUNT THE WATERING CAN

Use one of the 1/4" screws and one of the nuts to mount the watering can base to the deck.

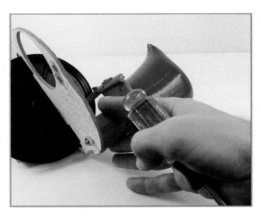

Figure 9-22 *Screwing watering can to the deck*

 Figure 9-22 was taken after the watering contraption had been painted red. I've included the photo, taken after the fact, to show how to screw it to the deck.

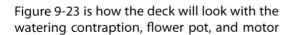

Figure 9-23 is how the deck will look with the watering contraption, flower pot, and motor wheel in place (you'll secure the motor later, after painting the deck).

Figure 9-23 *Deck with test parts in place*

ELECTRO-MECHANICAL PREP

To prep for building the electro-mechanical components of the Flower 'Bot, print the wheel nub (*fBot_wheelNub.stl*), then drill a 3/32" hole toward the outer edge of the motor wheel about 1/4" in from the wheel's outer ridge so the nub will clear it. See Figure 9-24.

Figure 9-24 *Drilling the motor wheel hole*

Figure 9-25 *Press-fit wheel nub*

Press-fit the nub into the wheel's nub hole (Figure 9-25).

The final 3D-printed part you'll need is the spacer for the soil moisture sensor probes (Figure 9-26).

Figure 9-26 *Probe spacer*

PAINTING

If you haven't printed your parts in their final colors, this is a good time to paint the robot.

Use a fine, white spray primer, such as Tamiya model primer or automotive primer (Figure 9-27). Follow the directions on the can and prime in a well-ventilated area.

Let the first coat dry, and prime a second time for best coverage.

You can then use acrylic craft paint and brushes to paint the robot in your favorite color scheme—even using techniques such as ink washes and dry-brushing highlights (see Figure 9-28).

Figure 9-27 *Priming feet*

Figure 9-28 *Painting feet*

There are many good resources on the Internet on painting models, so we won't go into details here.

Once the Flower 'Bot is painted and has dried, glue the legs, stovepipe, and railing in place. Then seal the paint with a matte finish spray sealer to prevent the paint from chipping (see Figure 9-29).

Figure 9-29 *Painted Chauncey*

ELECTRO-MECHANICAL AND SOFTWARE OVERVIEW

The Wrylon Robotical Flower Care Robot has a sophisticated flower care and watering system.

It's comprised of an Arduino brain ("Chauncey's Thought Process"), resistance moisture probe, motor driver shield, limit switches, test switch, and geared motor, as shown in Figure 9-30.

CHAUNCEY'S THOUGHT PROCESS

1. "Is the soil wet?"

2. "Yes? Good, I'll do nothing…"

3. "Oh, now it is dry? OK, I'll engage my motor in the forward direction so that it tips the watering can mechanism, thus watering the lovely little flower."

4. Repeat.

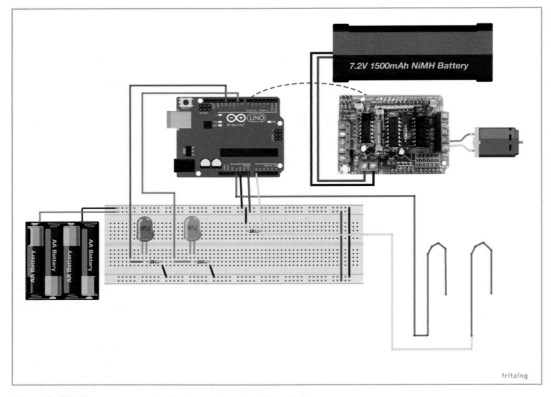

Figure 9-30 *Flower care and watering system breadboard diagram*

Here's how to put it all together.

INSTALL THE ARDUINO IDE

Begin by getting the Arduino board up and running. See the Arduino site (*http://bit.ly/ 1WwogWW*) for instructions on installing the Arduino IDE software on your computer. Make sure you can successfully upload code to the board.

 Disconnect the Arduino from USB or other power sources while connecting the other components.

INSTALL THE MOTOR SHIELD LIBRARY

Before you can run the *WrylonRoboticalFlowerBot.ino* Arduino sketch, you'll need to install the Adafruit Motor Shield v.2 library. Otherwise, you'll get error messages and won't be able to compile the program.

ADDING LIBRARIES

It's easy to add the Motor Shield library with the built-in Arduino library manager (Arduino Sketch menu→Include Library→Manage Libraries). Search for "Adafruit Motor Shield v.2," install, and restart Arduino.

If you need step-by-step help with adding libraries, see Appendix A for more information.

 Adafruit provides in-depth information about Motor Shield v.2 in its product tutorials (http://bit.ly/1O6vX4u).

CHAUNCEY'S CODE

Download the *WrylonRoboticalFlowerBot.ino* Arduino sketch (*http://bit.ly/1Gfkec6*) and look it over. The code is extensively commented to explain what each line does, but here's the general logic of how it works:

1. When the Arduino is powered on, it lights up the green LED, then checks the moisture probes to measure their resistance. If the reading indicates that there is adequate moisture (low resistance) in the soil, then all is good and nothing happens.

2. Every two seconds, it will check on the soil's moisture level. If, during one of these checks, the reading indicates that the soil is dry (high resistance), it will light up the red alert LED, and turn the motor for a few seconds to tip the watering can, thus moistening the soil.

3. It will continue taking readings, and once the level of moisture is sufficiently high, the motor and red alert LED will be turned off.

HARDWARE ASSEMBLY

Now that you know how the software will work, it's time to assemble the hardware. There are two options for making the electronics connections:

1. Wire all components directly to the ScrewShield or a prototyping shield.

2. Solder interconnects between the component wires and another set of matched wires on the ScrewShield or prototyping shield.

The interconnect approach enables easy part connection or disconnection, but it's an entirely optional extra step and will not impact the functions of the robot.

Connect your ScrewShield or prototyping shield to your Arduino.

POWER LED

Solder the 10K resistor to the ground leg (the short one) of the green 10mm LED (Figure 9-31).

This resistor will bring the current down to an acceptable level for a softly glowing indicator LED, as the direct power from an Arduino pin may damage a nonprotected LED.

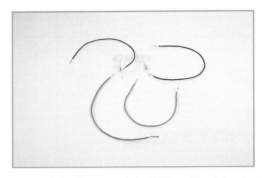

Figure 9-31 *Power and alert LEDs with soldered resistors*

 Different LEDs have their own power characteristics; the red LED will require less resistance; see "Alert LED".

Strip about 1/4" of insulation from both ends of a 10" length of black wire. Solder one end of the black wire to the free end of the 10K resistor.

Slide a piece of heat shrink tubing over the ground leg/resistor solder joint and another over the resistor/black wire solder joints, and then heat the tubing with the side of your soldering iron or heat gun. The heat shrink tubing acts as insulation from electrical shorts (Figure 9-33).

Figure 9-32 *Red LED being soldered to wires*

Strip the purple wire that will run the green LED to Arduino `PIN` 12. Then, solder it to the long leg of the green LED, and insulate the joint with heat shrink tubing.

The green LED assembly is now complete. If you plan to solder optional interconnects to its leads, you may do so now (Figure 9-34).

Connect the green LED's black ground wire to any ground (`GND`) terminal on the Screw-Shield or pin/pad on the prototyping shield.

Figure 9-33 *LEDs with heat shrink tubing*

 It's not recommended that you plug wires directly into female header pins on the Arduino as these connections tend to fall out. Chauncey takes no responsibility for the health of your flower should you choose to ignore this warning.

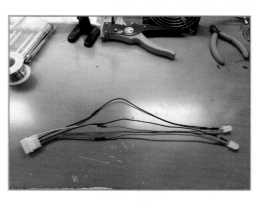
Figure 9-34 *Interconnects attached*

Connect the green LED's purple wire to the digital `PIN` 12 terminal on the ScrewShield or pad on the prototyping shield.

ALERT LED

For the red 10mm LED, repeat the same steps as with the green LED, except use the 220ohm resistor on the ground lead (short leg) and an orange wire for the positive lead (long leg). Again, solder on optional interconnects at this point if you like.

Connect the red LED's black ground wire to any ground (`GND`) terminal on the ScrewShield or pin/pad on the prototyping shield. Connect the red LED's orange wire to the digital `PIN` 9 terminal on the ScrewShield or pad on the prototyping shield.

MOISTURE SENSOR PROBES

The soil moisture sensor probes are two pieces of conductive metal with a wire connected to each of them.

The resistance between the two probes is nearly infinite in very dry soil, nonexistent when they are in direct contact with each other, and of varying values when they are both in contact with soil of varying moisture levels.

For moisture calibration to be meaningful, the probes must be kept at a consistent distance apart, so you will use the 3D-printed probe spacer and robot deck holes to maintain their distance.

OPTIONAL: BRASS TUBING

To create the stylish bends in the probes (an optional step you can ignore; everything will still work) you'll heat up the brass tubing before bending, otherwise the tubes will flatten out at the bend.

Using the 3″ stainless steel nails instead of the tubing will not impact Chauncey's performance. The brass tubing choice is purely aesthetic; the nails will work just as well.

If you're using the nails, you can skip down to the "Solder the Probe Wires" section.

HEAT AND BEND THE TUBING

In a safe place for an open flame, clamp the brass tube gently into the vise. Don safety goggles. Ignite your propane torch to a moderate flame.

Heat the tube at a point about 7″ up from one end, moving the flame back and forth over a 3″ section of the tube until hot enough to be bent, approximately 10 seconds (Figure 9-35).

Figure 9-35 *Heating the clamped tubing*

Use the pliers and bending form (e.g., a hammer head) to bend the brass into a U-shape at the heated point, as shown in Figure 9-36 and Figure 9-37. Let the tube cool before setting it on a surface or touching it.

Repeat for the second tube so they both have the same bend. Now that the brass tubes are bent, wait for them to cool down (see Figure 9-38).

Figure 9-36 *Bending the hot tubing*

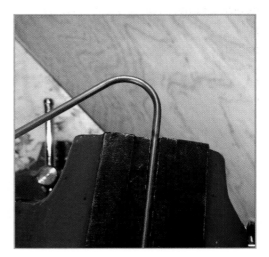

Figure 9-37 *Creating the "U" shape*

Figure 9-38 *Bent tubing*

SOLDER THE PROBE WIRES

Next you'll solder wires to the ends of brass tubing or the stainless steel nails. Strip 1" of insulation off of the ends of a 10" length of blue wire.

Wrap one end of the blue wire around the brass probe's long end, and then solder it in place (Figure 9-39).

Wait a moment for the end to cool, then slip a piece of heat shrink tubing over the solder joint and heat it up to shrink.

Repeat the stripping, soldering, and heat shrink tubing steps for the second probe, using a length of yellow wire (Figure 9-40).

Figure 9-39 *Wrap and solder blue wire*

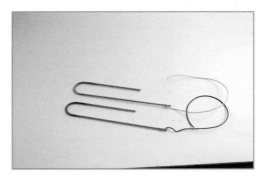

Figure 9-40 *Tubes with soldered wires*

MOISTURE SENSOR CIRCUIT

In order for analog `PIN 0` on the Arduino to read the resistance between the probes, it needs a stable reference point. This is done by running one of the probes to 5V on the Arduino.

The other probe wire will connect to a Y-adapter (see Figure 9-41) that runs one end to GND through a resistor and the other end to analog `PIN 0`.

Assemble the Y-adapter

Cut another short length (around 6") of yellow wire. Solder it to the free end of the existing yellow wire you previously soldered to a brass probe.

As seen in Figure 9-42 solder one leg of a 10K resistor to this junction between the two yellow wires.

Solder a 6" length of black wire to the free leg of the 10K resistor you just soldered into the Y-junction.

Optionally, you may add interconnects for the probes at this point.

Connect the probes to the shield

Connect the blue probe wire (either directly or through the interconnects) to 5V on the ScrewShield or prototyping shield (Figure 9-43).

Connect the yellow probe wire to the analog `PIN 0` terminal on the ScrewShield or pad on the prototyping shield.

Connect the black probe/resistor wire to any GND terminal on the ScrewShield or pad on the prototyping shield.

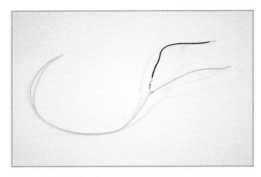

Figure 9-41 *Y-adapter*

Figure 9-42 *Assembling the Y-adapter*

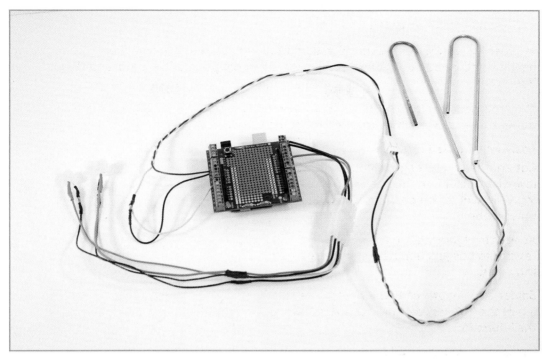

Figure 9-43 *Probes connected to shield*

THE MOTOR

The Arduino doesn't have the ability to directly control a strong motor; for that you'll add the Motor Shield. The Motor Shield is able to use a bigger power supply, separate from the Arduino's main supply. This enables the Arduino to control motors that draw high current.

Place the Motor Shield on top of the ScrewShield or prototyping shield that is already on the Arduino (this is called "shield stacking"); see Figure 9-44.

Screw the motor's two wires each into their own terminal of the M3 block of the Motor Shield. You may optionally choose to use an interconnect here, instead of direct wiring.

Remove the Motor Shield power jumper, if it's in place. This will force the shield to only draw power from the external power supply, not from the Arduino.

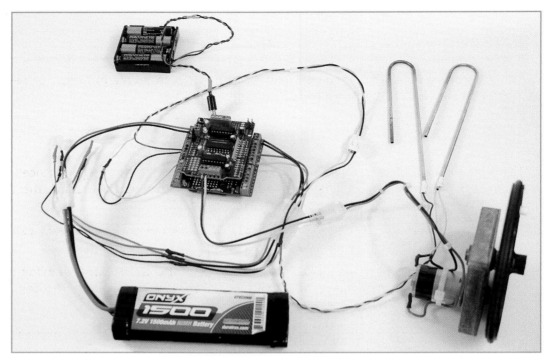

Figure 9-44 *Adding stacked shields and motor*

Connect the motor battery positive wire (usually red) to the Motor Shield positive (+) power terminal, and the battery's ground wire (usually black) to the Motor Shield ground (-) terminal.

You should see the Motor Shield green LED light up.

 Check power and ground with a multimeter if you're at all unsure, as you can damage the electronics if you switch the polarity.

INTERCONNECT OPTIONS

Again, you may choose to use an interconnect for the motor battery, in order to more easily facilitate occasional recharging of the battery. One option is to use pliers to pull two wires and female pins from an old four-pin Molex hard drive connector, which may fit nicely with the interconnect already in place on an RC car battery.

PROGRAMMING THE ARDUINO

With the hardware built, it's time to upload the sketch (program) to the Arduino. But before you do, take a look at the sketch you downloaded previously.

NOTABLE VARIABLES

There are a couple of variables to pay attention to at the top of the program. Since different flowers may require different moisture levels, you may want to adjust the `dryValue` variable. A higher number will lead to moister soil.

Once you've uploaded the code and put the probes in the flower pot, you'll be able to test a well-watered flower to get a baseline value.

The other variable to check out is `motorRunValue`. This number represents how many seconds the motor will run in order to activate the watering can mechanism. The default is 8000 milliseconds (eight seconds).

UPLOAD THE CODE

1. Plug the Arduino into the USB cable connected to your computer, and plug the motor battery into the Motor Shield if it is not already.

2. Set the Arduino IDE board type to "Arduino UNO," Tools menu→Board→Arduino Uno.

3. Pick the proper serial port in the Tools→Serial Port menu. Then upload the *Wrylon-RoboticalFlowerBot.ino* sketch to the Arduino.

PLACE THE ELECTRONICS

Figure 9-45 *Motor and wheel seated in deck*

Seat the motor and wheel in the proper place on the deck, and secure the motor with the two screws and bolts (Figure 9-45).

Place the flower pot in its hole in the deck.

Hook the two probes through their deck holes (Figure 9-46).

Figure 9-46 *Probes through deck holes*

With the moisture sensor probes spaced apart properly with the 3D-printed sensor spacer, place the probes into the soil of the flower pot, about 2-1/2" deep (Figure 9-47).

Figure 9-47 *Probes in soil with spacer*

MONITORING MOISTURE

During your initial testing, it is advisable to connect Chauncey's Arduino port via USB cable to your computer and monitor the moisture levels.

Open the serial monitor (Tools→Serial Monitor) in the Arduino IDE, and you'll be able to view the sensor `moistValue`.

The number being reported for a recently watered plant is a good baseline number to know, but more important is to note the value when the soil is dry enough that you want Chauncey to engage the watering routine.

This is the number you will set in the sketch's `dryValue` variable. You can then resave the sketch and upload it again to your Arduino.

BRAIN IMPLANT AND ASSEMBLY

The electronics, power, and motor of the Flower 'Bot are ready to be implanted.

First, close the Arduino IDE serial monitor if it is open, and disconnect the Arduino from your computer USB cable.

Next, place the Arduino stack in a plastic bag inside the Flower 'Bot's belly (Figure 9-48). This will prevent any stray drips of water from shorting the electronics.

Figure 9-48 *Bagged electronics*

Push the green and red 10mm LEDs up through the holes in the deck. They should fit snugly, but you may also use a bit of glue to secure them, if you like. The LEDs can be seen in Figure 9-50.

Make sure the motor battery is connected, then place it inside the belly of the 'Bot.

Connect the 4-AA battery pack plug to the Arduino (this will power it on) and then put the deck on the Flower 'Bot (Figure 9-49). Implant complete!

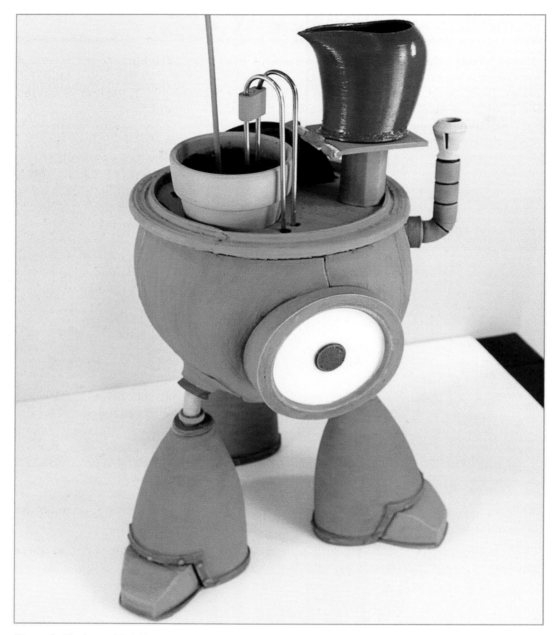

Figure 9-49 *Assembled Chauncey*

TESTING/DEPLOYMENT

Using a small pitcher, fill Chauncey's watering can with water. When the soil is dry enough to cause the watering sequence, the red alert LED will light up and the motor wheel will begin to turn, making contact with the watering lever (Figure 9-50). As the lever is pushed, the watering can will tip toward the flower pot, and pour in its water.

Figure 9-50 *Watering lever moves after detection of dry soil*

You may now leave your flower in the care of the Wrylon Robotical Flower Care Robot, secure in the knowledge that it shall be well cared for, always (Figure 9-51).

Check the watering can periodically to see if it needs refilling, should the watering operation take place while you are away.

Figure 9-51 *Watering operation: can at rest, tilting up, watering flower*

UPGRADES

Since the mysterious closure of Wrylon Robotical Industries, it is no longer possible to send your 'Bot in for upgrades. However, don't despair, you may further customize your Flower 'Bot yourself.

Stylistically, you could begin by building a wooden deck (cut 1/8" Baltic Birch plywood on a laser cutter or jigsaw with the included file, *fBot_xx_woodDeck.svg*, and 3D-print the accompanying motor bracket, *fBot_xx_motorBracket.stl*, to screw into the wooden deck), creating brass rails by bending and soldering more 1/8" tubing, or even sinking heated brass nails into the body of the 'Bot to simulate metal rivets.

Other possible upgrades could include:

- Adding connectivity with an Ethernet shield and modifying the software to send email alerts, or adding an SMS shield to send text messages

- Giving Chauncey a chance to be heard with a sound wave shield or voice module and speaker for sound alerts/spoken updates

- Improving Chauncey's senses with a thermocouple to log temperature or even a light sensor to measure sunlight and motorized casters in his feet, allowing Chauncey to rotate during the day to follow the sunlight

With a FLORABOT 3L-1G, the only limitation is your imagination.

Figure 9-52 *Chauncey ventures outside*

Figure 9-53 *Completed FLORABOT 3L 1G production unit*

Appendix A: Installing Arduino Libraries

Before you can run an Arduino sketch that's dependant on external libaries, you'll need to download and install those libraries. Otherwise, you'll get error messages and won't be able to compile the program.

Installing libraries used to be a little more complicated, but it's easy now that the Arduino IDE has included a library manager in version 1.6.2 that includes all of the Adafruit libraries used in this book and many, many more.

STEP 1: ACCESS THE LIBRARY MANAGER

To add a library, access the Library Manager by selecting the Arduino Sketch menu→Include Library→Manage Libraries. A new window will open listing the Arduino reference libraries (Figure A-1).

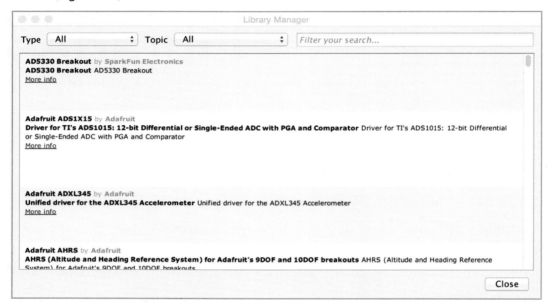

Figure A-1 *Arduino Library Manager*

STEP 2: SEARCH

Search for the library you're looking for to bring it to the top of the list. For example, search for "adafruit motor shield v2," click the library to select it, and an "Install" button will appear (Figure A-2).

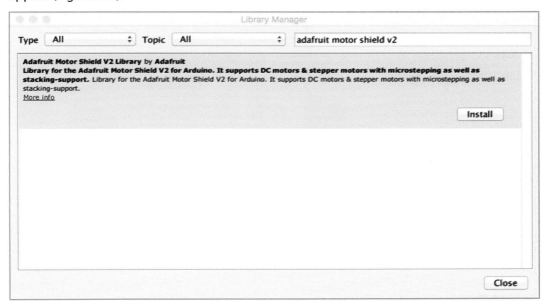

Figure A-2 *Install library*

STEP 3: INSTALL AND RESTART

Install the libary and restart Arduino. If you go into the Library Manager and search again, you'll see that the library has been added (Figure A-3).

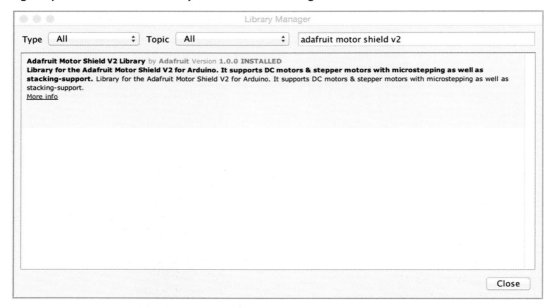

Figure A-3 *Library installed*

GOING FURTHER

The Arduino Library Manager is great for installing the necessities. However, if you find yourself off the beaten path and need to install libraries the old-fashioned way, or just want to learn more, check out these resources from Arduino (*http://bit.ly/1MZ011I*) and Adafruit (*http://bit.ly/arduino-libs*).

Index

V

VCP drivers, 70
video stabilization (see camera gimbal)

W

wheel assembly, 152-154, 157-159

wooden chassis, 80-82

Y

Y-adapter, 242